SOTHIS PRESS

Sothis Press aims to completely reimagine the public domain book with an emphasis on universal access to captivatingly designed print editions. We believe that technology will enable every single book ever published to be perpetually accessible in beautiful print form. Sothis is beauty in ubiquity.

www.sothis.press
1 Waterhouse Square, London, ECIN 2ST

A Treatise on Hat-Making and Felting

Including a Full Exposition of the Singular Properties of Fur, Wool, and Hair

John Thomson

SOTHIS PRESS

ISBN 978-1-0368-0116-8

A Treatise on Hat-Making and Felting
is in the public domain.

This edition is published by
Sothis Press © 2023. The company
is registered in England and Wales.
The Company Number is 14900389.

HAT-MAKING AND FELTING,

INCLUDING A FULL

EXPOSITION OF THE SINGULAR PROPERTIES
OF FUR, WOOL, AND HAIR.

BY
JOHN THOMSON,
A PRACTICAL HATTER.

PHILADELPHIA:
HENRY CAREY BAIRD,
INDUSTRIAL PUBLISHER,
406 Walnut Street.
LONDON:
E. & F. N. SPON,
48 Charing Cross.
1868.

PHILADELPHIA:
COLLINS, PRINTER, 705 JAYNE STREET.

CONTENTS.

TREATISE ON HAT-MAKING AND FELTING.

IT is conceded as an axiom, that theory and practice, in the pursuit of any object, are in their natures essentially different and distinct. But at the same time they long for a mutual understanding each to confirm the assertions of the other, the consummation of all practical results being the mutual embrace and perfeet reconciliation of these two attributes.

The writer of these pages, being a practical hatter, desires to describe intelligibly his calling, dispensing with all technical terms, at the same time conscious of being liable to receive an unfair criticism from his brother tradesmen, although perfectly innocent on their part, resulting from the prejudices engendered by the many would-be secrets that pertain to the different work-shops, together with their various modes and methods of working, all of which most generally are but trifles merely to gain a name.

The practice of a trade without a knowledge of the why and the wherefore of certain usages is a sad defect in any workman, but more especially in certain trades: Hatting being one of those which depends upon *second* causes for its proficiency, we venture here an explanation with perfect confidence, hoping that the fraternity of hatters will be indulgent, and that they may profit by an experience of many years in the trade, and that for one error or omission in the writing of these sheets they will find compensation in the new ideas that will spring from their perusal, which may be an incentive to further improvements in the business resulting beneficially to all.

Theory without practice, or practice without theory, is like groping in the dark, and perfection in no trade can be attained till every effect can be traced to its cause, and *vice versa.*

It is much to be regretted that practical operative workmen are so diffident in writing and publishing their experience in their several trades and occupations, quietly permitting theorists ignorant of the business to glean as best they can from other parties the most intricate and complicated particulars of a trade, and hence the attempt to illustrate the most useful branches of an art often results in crude and even erroneous descriptions of things of the greatest moment, and the dissemination as correct, that which is altogether at variance with the truth. In confirmation of the above, we may instance the manufacture of hats as described in a work of much merit, and which is accounted as worthy of all confidence, wherein the error above spoken of is but too plainly visible. Thus, in the supplement to the third edition of that most respectable work the Edinburgh "Encyclopedia Britannica," in the article Hat, an apology is made for the original treatise upon that subject, it being acknowledged as both defective and erroneous from the imperfect source of the information. Such a confession, and from such a source, sufficiently exonerates any one from egotism in an attempt to write a more perfect and correct description, coupling theory with practice; relieving the felting process from its misty obscurity by a faithful expose of the whole system : well knowing that an increase of business, like free trade, will be the result of a right understanding of a formerly supposed mystery, viz., the True cause of Felting.

Felt and felted articles being already in use, in many trades in addition to that of hat-making, necessitates a general and indeed a very full and lucid description of the materials of which they are made.

Descriptions of Furs, Wools, Hair, &c.

Fur, properly speaking, signifies the skins of various species of animals, dressed in alum or some other preparation with the hair on, and made into articles of wearing apparel ; but the term fur also signifies the stuff that is cut from the skin, for the use of the hatter, and in this sense alone it will be employed in the following pages.

Hair, wool, fur, and animal down are simply slender filaments or thread-like fibres issuing out of the pores of the skins of animals, and all partaking of the same general nature, such as great ductility, flexibility, elasticity, and tenacity, differing entirely from the vegetable wools and downs, such as cotton, &c., which contain neither of these four great characteristics to any valuable or appreciable extent.

To characterize in a familiar way these several grades of material, it may be said that fur is distinguished from wool by its greater fineness and softness, and hair from wool by its straightness and stiffness. The nature of all these bearing some relation to each other, it will be necessary in this treatise to use the word hair occasionally to designate one and all of them, that word being most convenient, and tending to avoid confusion.

Simple as the idea may be, and though trifling in appearance, yet the study of a single hair is particularly interesting, both to the naturalist and the man of business, as will be seen when we mention a few of its many peculiarities; hoping it will prove a source of enjoyment to the one and a profit to the other.

Hair, wool, fur, &c., form quite an extraneous appendage to the skin, or body producing them, not at all directly dependent on the life of the animal for their own existence, for they have been known to live and grow for some time after the death of the animal itself. We also know that they live, grow, and die, showing all the signs of youth, maturity, and old age. Hair possesses no sensation at any period of its existence; of itself it has no feeling of touch, nor has it the power of voluntary action.

The growth of hair is peculiar as it projects and grows in length from the root, and not by the top as with vegetable productions, the lower portion lengthens out, and the top is merely projected forward; and when once cut, it never again resumes its tapering point.

Hair or fur of whatever quality, consists of a single slender filament, without a branch or knot of any kind, and that filament is a tube, which is filled with a fat oil, the color of the hair being derived from this oil.

By the chemical analysis of hair it is found to consist of nine different substances: 1st, gelatine or animal matter, which constitutes its greater part; 2d, a white concrete oil in small quantity; 3d, another oil of a grayish-green color more abundant, these oils comprising about one-fourth of the entire weight; 4th, a few particles of oxide of manganese; 5th, iron, the state of which in the hair is unknown; 6th, phosphate of lime;

7th, carbonate of lime in very small quantity; 8th, silex in greater abundance; 9th, and lastly, a considerable amount of sulphur—such is the constitution of all furs, wools, hair, &c., most of which may be dissolved in pure water heated to a temperature above 230° of Fahrenheit, by which it is partially decomposed. Hair is likewise soluble in alkalies, with which it forms soap. Chlorine gas immediately decomposes it, producing a viscid mass.

It is worthy of particular remark, that of all animal products, hair is the one least liable to spontaneous change, evidence of which may be found in the fact that the Peruvian, Mexican, and Brazilian mummy hair is still perfect, and is supposed to be from 2500 to 3000 years old, and stands the hygrometric test with equal firmness. From this we should suppose the body or substance of hair and wool to be exceedingly hard and solid, which is really the case, as no pressure has yet been applied sufficiently powerful to entirely deprive wool of the water with which it has been washed—the interstices between the fibres of the assemblage never having been closed by the power applied, as the water therein collected may still be drained off when the pressure is removed.

Although hair is of a tubular construction, yet all varieties are not of a completely cylindrical form; a curl is the result of all flat-sided or oval hairs, the exceeding oval being the unfailing characteristic of the negro race. A cross section of a hair, if circular, denotes the long, soft, and lank fibre of a cold northern animal; but if the cross section shows an extreme flat-sided hair, that hair will be crisp and frizzled, and of a tropical extraction. Quite a gradual change in the form of the fibre of hair is observed in all animals as we ascend from the equator to the highest latitudes, other things being equal.

It has long been a desideratum how to discriminate between the various qualities of hatters' *fine* furs, and no really reliable test has yet been obtained, superior to the judgment of the human eye, the fineness of fibre for the hatter being of most essential importance, particularly that allotted for the flowing nap upon the outside of the hat. Although the thickness of the fibre of the finer furs has never been properly gauged, it will be a source of some satisfaction to know that the diameter of the human hair varies from the 250th to the 600th part of an inch, while the fibre of the coarsest wool is about the 500th and the finest about the 1500th part of an inch.

Hair may be bleached on the grass like linen, after previous washing and steeping in a bleaching liquid, after which it may be dyed of any color.

It is very doubtful whether the growth of hair can by any artificial means be expedited, or the hair itself increased in length, in quality, or in density. A fine field of enterprise would be opened for the fortunate inventor who could increase the produce of the finer and more expensive furs. In contradistinction to this, however, it may be stated that the inhabitants of some countries, the Malays, for instance, purposely destroy their hair by using quick-lime.

We come next to describe minutely another peculiarity appertaining to hair, upon which all felting or shrinking of a fabric depends; that grand secret that has been a mystery in all ages, until within a few years, or at best was only surmised. Upon this property alone depends the whole art of hatting and of felt making, whether in sheets or otherwise, as well as the fulling of cloth and the shrinking of flannels, and all articles the material of which is made of wool, hair, or fur.

As many branches of business depend for their success upon the *non-shrinking* quality of their goods, a study of the felting principle becomes quite appropriate and interesting to those manufacturers, whilst perusing that of the opposite. Pulled wools, rather than cut or shorn wools, must always have the preference with the one class of manufacturers; at the same time, the other class must adhere tenaciously to those which have been cut, the roots of the hair causing all the difference, for that remarkable quality, the felting principle, is upon all the same whether pulled or cut.

A few familiar facts dependent upon this inherent felting quality of hair will aid the illustration. When a hair is held by the top, it can be severed with a razor much more readily then if held by the root. Again, a hair held by the root, and drawn through between the finger and thumb, feels quite smooth, but when held by the top, a rough and tremulous motion is perceived.

Again, place a hair of three or four inches in length by the middle, between the finger and thumb, and twirl it a few times, when the hair will be found to proceed towards one end, as the twirling and rubbing are continued, and invariably advancing root end foremost, whichever way the hair is placed between the fingers. If two hairs are used in this example, lay the root of the one to the top of the other, their respective motions will be doubly discernible.

The cause of all these singularities of the hair it is now designed to explain, which shall be done as explicitly and concisely as possible, with a few proofs of its astonishing power in a collective capacity.

The above-mentioned phenomena are the result of that same long-hidden property, and which is nothing more than a certain clothing or covering, entirely surrounding the stem of every hair, in the form of very minute scales, so very minute, indeed, that it requires the aid of a very powerful microscope to enable the beholder to discern them, and even then but faintly. These scales, which cover thickly every filament of animal hair, wool, fur, &c., are thin pointed lamina, quite similar to the scales on a fish, and overlapping each other as do the shingles or slates upon a house. This state of the hair being understood, the *modus operandi* of the above examples may be thus explained: When the hair was held by the point, it was easily cut by the edge of the razor entering under the scales; but when held by the root, the instrument slipped smoothly over them; and the hair that was drawn through the fingers, when held by the point, felt rough and tremulous, from the jagged points of the scales, but smooth when drawn in their own direction.

The twirling of the hairs between the finger and thumb, resulting in their travelling motion, was on account of the points of the scales catching on the fingers, in the act of rubbing, similar to the heads of wheat or barley at harvest time which school-boys put into the sleeves of their coats, and which are sure to come out at some other extremity to that at which they were put in, caused by the working of the boy's arm upon the jaggy beard or awn of the barley head.

The task of counting the number these lamina that clothe the body of these hairs, must have been both tedious and difficult, from their very minuteness and profusion. On a single filament of merino wool, as many as 2400 barbed scales, like teeth, projecting from the centre stem, have been counted in the space of one inch. On Saxony wool there were 2700, while other wools were as low as 1860, and none were found to have so few as 1000 to the inch.

No vegetable wools whatever, such as cotton, &c., have any such appendage upon their fibres, and, consequently, cotton or cotton goods never shrink in the act of washing, as woollen goods do. Cotton, therefore, never can become a suitable material for felting purposes, every fibre being smooth from end to end in either direction, and in contradistinction to fur, which, though equally smooth as the cotton in one way, rebels triumphantly when irritated in the contrary direction, as already described. Mechanically speaking, cotton is smooth, solid, and triangular, whilst wool is rough, tubular, and cylindrical.

The grand cause of that mysterious and curious operation called felting, fulling, shrinking, thickening, and solidifying of a fabric, whether of original loose wool, fur, or other stuff, or of that spun into yarn and woven into cloth, is the presence of these scales.

Till lately, the best operative hatter and the investigating philosopher were equally at a loss to explain upon what principle such effects were produced. Take, for instance, a handful of wet fur or wool, which is merely an assemblage of hairs; squeeze and press it, work it a little in the hand, and then observe the effect; for immediately upon pressing it a certain locomotion is thereby conferred upon every fibre of that assemblage, which is increased by every turn of position that is given to the body of wool. The rolling and pressing change the position of each fibre.

A friction is produced upon every member composing the mass; a footing as it were is obtained from the scales of each, and the fur or wool being all bent or curled, a progressive motion goes on, interlacing each other in their travels, resulting in a compact, dense body, which may well challenge the goddesses of both patience and perseverance to undo. Every hair has been travelling in its own individual direction, boring, warping, grasping, holding, and twisting amongst its fellows like a collection of live worms.

The power of combination, like the fable of the bundle of sticks, is strikingly illustrated in the case of the hair, which when viewed singly seems so very insignificant, but collectively, and when pressed by the hand of oppression, hardship, and ill treatment, they combine and become strong and defiant, clasping each other in their embrace, tenaciously clinging to each other the more they are tortured, as if they were living rational beings, conscious of their innocence, and free from guilt.

Stockings, for instance, that are knit with soft-spun wool, for the use of whale fishermen in northern latitudes, are large enough, when first formed, to hold the whole man, but are felted down to the required size in the fulling mill, where they are battered, tossed about, and tortured to that degree that is required by their tormentors. The writer has seen a millful of these stockings whose sides were felted so firmly together, from a neglect of the workmen to turn them inside out, in due time, during the felting operation, that a knife was required to open them, and which actually failed in several instances, so firmly had their two sides grown together; common tearing having no effect whatever, each and every single hair had embraced its neighbors, and their mutual action defied all attempts to open these stockings.*

There are instances of ruminating animals having died from the effect of balls of hair having formed within their stomachs, hair by hair having accumulated while licking themselves with their tongues. These balls are all found to be as perfectly felted as the natural bend of the several hairs composing them would allow, the felting having been accomplished by the motions of the intestines of the animals. The disgorged balls from the stomachs of nocturnal fowls are all of the same nature.

As has been said, felt may be made of any kind of animal fur, wool, or hair, provided it be bent, crimped, or curled, for if straight as a bristle it would work out of the mass as readily as into it, and lose itself in the operator's hands.

All materials intended for felting must be cut from the pelt or skin, and not pulled, for the obvious reason

* The most familiar instance of mutual association and combination, resulting in real utility, though not so striking on account of our familiarity with it, is the broad-cloth of which our clothes are made, which when cut by the tailor will never unravel. This result is wholly the effect of its felting in the fulling mill during the operation of scouring and washing, every fibre of the wool of which the cloth is made, having clung to its immediate neighbors, both warp and weft, and with the spirit of true friendship they still remain in each other's embrace, and the cloth is transformed from a loose to a solid fabric.

Another instance of the power of combination is the mysterious Gordian knot that we read of in history, which promised the empire of the world to him who could unloose it, and which Alexander the Great is reported to have cut with his sword, because he failed in the attempt. If not a fabulous story, that compound knot the illustrious Gordius is supposed to have cunningly felted previous to hanging it up in the temple.

that a pulled hair invariably brings with it its root, in the form of a button or bulb, which would greatly impede its progressive motion in the act of working, as each and every hair under the operation of felting bores into and amongst the other filaments of the fur composing the mass, root end foremost, a sharp point therefore is obtained by cutting. This rule is universally and invariably adopted by all hat furriers.

Wool of any great length of staple, after being carded, is pressed, and either clipped, cut, or chopped into shorter lengths, which facilitates the felting operation, and improves the solidity of the felt that is produced.*

The various materials most used in hat-making are the furs of the beaver, the otter, the rabbit, the hare, a species of the muskrat, a species of the monkey, a species of the seal, and a few others, together with Saxony and Spanish wools and the hair of camels and goats. Numerous as are these various names, most of the animals produce five or six different qualities of stuff, from particular parts of the same skin, varying greatly in price or value.

The finest furs all come from those animals that inhabit the coldest climates, and the season of the year in which any of them are killed greatly influences the quality of the fur; a summer skin of some of these animals being comparatively valueless, however excellent it might be in the winter season. And what is particularly worthy of the hatter's attention

* The reason why wool and woollen goods felt and solidify more readily than any straight fibred furs, is owing to the natural curl or frizzle possessed by wool, each and every bend of every individual filament assuming an inclination for travel independent of each other and of the general inclination of the perfect fibre.

is, that fur that has been kept one or two years, after being cut from the skin, produces a better working, and a more solid article of felt, than fur from a newly-killed animal. The lamina of such fur seem to rise and erect themselves upon the stem of the hair by being kept, which may account for its better felting quality. This would appear to be confirmed by the well-known fact that the 5 lb. bags in which old fur stuffs have been kept are generally burst open.

One or two properties peculiar to furs and wools may still be mentioned, as, for instance, all felting, by whatever means accomplished, necessitates either a damp or wet process with the aid of heat, and the facility of thickening or solidifying is accelerated by the application of soap to the material under the operation. Or the water may be acidulated for the same purpose with a little sulphuric acid, as either of these acts as a penetrating solvent upon the natural oil of the animal which still remains between the stem and lamina or scales of the hair, thus baring the barbed points of the crusty scales, the better to catch and hold their grip upon each other.

Oil or grease, on the contrary, when applied directly upon wool, covers up these lamina or scales, thereby destroying their felting power, as is well known to all wool spinners, however little they may understand the real cause of its being so, further than the fact of giving to it a smooth gliding effect, so necessary for the object of their business.

It may be amusing, whether true or not, to know that the rude Turcomans are said to dwell, even to this day, in tents covered with felt, which they make by treading with their feet the raw material of which it is made, whilst it lies upon the ground, thus favoring the supposition that felting was invented prior to weaving.

However, so far as we can learn, a real systematic method of felting is comparatively of a late date, and until within a few years felt has been chiefly employed for hats and hats alone. This is, however, now but a branch of the felt manufacture, for plaids, coats, vests, pants, leggings, shoes, gaiters, slippers, mittens, and caps, the covering of steam cylinders and boilers, carpets, polishing cushions for jewellers and marble cutters, covering for the roofs of houses which is afterwards waterproofed, as also linings of water-tight compartments in ships and ship sheathing, and the covering for the blocks of calico and other printing, &c. &c., are now made of this material. As the nature of hair and the principle upon which its felting property depends become better known, the manufacture of felt will be stimulated and increased, and applied to many purposes other than those above enumerated, and not imagined at the present time.

The high price of the finer furs, resulting from the indiscriminate destruction of the animals which produce them, forms the only apology for the introduction of an inferior material into the body of the manufactured article. Cotton, which is of quite a limber nature, is too pliable, as indeed are all vegetable products when mixed with fur. They lie dead within the body of the mass, and if the labor be continued beyond a certain time, the active principle of the fur will be seen to have clung to itself, leaving the cotton quite exposed on the outside. Even under the most perfect manipulation, a mixture of cotton, from its want of elasticity, will give a product which is to a corresponding degree deteriorated.

There was a time when beaver skins were bought from the natives, by the Hudson Bay Company, at the regular price of 14 skins for a gun, 7 for a pistol, 2 for a shirt or one pair of stockings, 1 for a comb, or twelve needles, &c. &c., less than the hundredth part of their real value, and all the other fur-bearing skins belonging to that country were rated by that of the beaver.

"The Scientific American" of New York for Dec. 1859 says that, not much more than half a century ago, not a pound of fine wool was raised in the United States, in Great Britian, or in any other country except Spain. In the latter country the flocks were owned exclusively by the nobility, or by the Crown. In 1794, a small flock was sent to the Elector of Saxony, as a present from the King of Spain, whence came the entire product of Saxony wool now of such immense value. In 1809, during the second invasion of Spain by the French, some of the valuable crown flocks were sold to raise money. The American Consul at Lisbon, Mr. Jarvis, purchased fourteen hundred head, and sent them to this country. A portion of the pure unmixed merino blood of these flocks is to be found in Vermont at this time. Such was the origin of the immense flocks of fine woolled sheep in the United States.

The same authority further adds that the simplest and most easy method of judging of the quality of wools, is, to take a lock from a sheep's back and place it upon an inch rule; if you can count from 30 to 33 of the spirals or folds in the space of an inch, it equals in quality the finest Saxony wool grown. Of course as the number of spirals to the inch diminishes, the quality of the wool becomes relatively inferior.

Cotswold wool, and some other inferior wools, do not measure more than nine spirals to the inch.

The Fulling Mill.

Having alluded to the fulling mill as a felting machine, it is only necessary to remark here, that it is a rude looking but effective method of condensing a *previously formed article.* It consists of a trough six or eight feet long, and two feet wide, varying in size according to the kind of goods to be operated on. The bottom is of a semi-circular form, having a radius of five or six feet, with sides rising three or four feet high, a strong solid heading, but no end piece. There is a heavy wooden battering-ram suspended from above, at a height answering to the curve of the trough; its immense head has a flat face fitting the trough in which it is made to play freely, similar to the pendulum of a clock. The goods are tumbled promiscuously into this trough in front of the ram, with warm water, fuller's earth, and soap, in sufficient quantity to saturate and wash the material, a small stream of water from a boiler being admitted for that purpose. The power of a water-wheel or steam engine draws back the ram out of its perpendicular, to its allotted distance, whence it falls by its own gravity, with a momentum that sweeps the goods before it with a fearful crash, upon the solid heading of the trough. On the withdrawal of this enormous hammer for a second onset, the goods roll over, resuming their quiescent state, but differently disposed, which is no sooner done than back comes the ram, repeating its dashing blows upon its unoffending and unresisting victims in the trough, washing, scouring, and buffeting them about, till they become not only clean, but completely felted.

All of our broadcloths have been subjected to its action, in the process of which the hairs of the weft and those of the warp have become mutually entangled, and each with one another, as with hatting in the regular hand process of loose wool or fur felting. Indeed, every hair composing the whole piece of cloth has its individual and independent progressive motions, combining the threads of both warp and weft together to such an extent that these cloths never unravel, and no hemming of a garment is required in the making of our clothes.

Twelve hours in the mill will reduce a piece of cloth two-fifths of its breadth and one-third of its length.

The progressive travelling motion of the hair resulting in the entanglement of the fibres and consequent felting and shrinking of the cloth, is further exemplified in the comfortable soft half-dress caps of the British soldiers, and in the bonnets and caps of the Scottish peasantry generally, which have all been first knitted very large with soft spun yarn, and afterwards felted down to the required size in the fulling mill. During the fulling of any and all kinds of goods, they must be frequently taken out of the trough, to be stretched, turned, the folds straightened, and generally inspected.

History of Hats and Hatting.

The word hat is of Saxon derivation, being the name of a well-known piece of dress worn upon the head by both sexes, but principally by the men, as a covering from the hot sun of summer, the cold of winter, a defence from the blows of battle, or for fashion. Being the most conspicuous article of dress, and surmounting all the rest, it has often been ornamented with showy plumes, and jewels, and with bands of gold, silver, &c. It is generally distinguished from a cap by its having a brim, which a cap has not, although there are exceptions even to this rule of distinction, for there are hats that have no brims, and there are also caps that are provided with a margin. Those hats that are made of fur or wool have all been felted, and felt strictly speaking is a fabric manufactured by matting the fibres together, without the preliminary operation of either spinning or of weaving.

We find but little of hat-making recorded in history, and anything relating to hats is extremely meagre, although their partial use may be traced back to the time of ancient Greece amongst the Dorian tribes, probably as early as the age of Homer, when they were worn, although only by the better class of citizens when on a distant journey. The same custom prevailed among the Athenians, as is evident from some of the equestrian figures in the Elgin Marbles.

The Romans used a bonnet or cap at their sacrifices and festivals, but on a journey the hat with a brim was adopted. In the middle ages the bonnet or cap with a front was in use among the laity, while the ecclesiastics wore hoods, or cowls.

Pope Innocent, in the thirteenth century, allowed the cardinals the use of scarlet hats, and about the year 1440, the use of hats by persons on a journey appears to have been introduced into France, and soon after became common in that country, whence probably it spread to the other European States.

When Charles VII. of France made his triumphant entry into Rouen in 1440, he wore a felted hat.

Hatters of the present day most generously ascribe the honor of the invention of felting, and of its prospective introduction to that of hat-making, to the old renowned Monk St. Clement, who when marching at the head of his pilgrim army obtained some sheep's wool to put between the soles of his feet and the sandals that he wore, which of course became matted into a solid piece. The old gentleman, philosophizing upon this circumstance, promulgated the idea of its future usefulness, and thus it is said arose the systematic art of felting and of hat-making.

However all this may be, still the invention of felted fabrics for the use of man may have been, as some assert, very ancient and of quite uncertain origin. The simplicity of its make, as compared with that of woven cloth, shows all speculative assertions to be rather uncertain.

However obscure the origin may be, we learn that the first authentic accounts of hatters appeared in the middle ages, in Nuremburg in 1360, in France in 1380, in Bavaria in 1401, and in London in 1510.

The hatting trade of the United States of America is noticed first in the representations made by the London Board of Trade to the House of Commons in the year 1732, in which they refer to the complaints of the London hatters, regarding the extent to which their particular manufacture was being carried at that time in New York and in the New England States.

The Fashions.

A look at the fashions and mode of dressing in ancient times causes amusement. So capricious is the fancy of man that nothing is immutable, all is change, and hats have been of all conceivable shapes and colors, and dressed with the most fanciful decorations, plumes, jewels, silk-loops, rosettes, badges, gold and silver bands and loops, &c. &c.

The crowns and brims having been in all possible styles from the earliest period. It would appear that nothing is left for the present and all coming time, but the revival of what has already been, even to the fantastical peaked crown that rose half a yard above the wearer's head.

In the fifteenth century, hats in Great Britain were called vanities, and were all imported, costing twenty, thirty, and forty English shillings apiece, which were large sums of money at that early period.

The most extreme broad brims were worn about the year 1700, shortly after which the three-cornered cocked hat came in, and about this time feathers ceased to be worn, the lingering remains being left for the badge of servitude to the gentleman's attendant. Metal bands and loops were only regarded as proper for naval and military men of honor.

It is a singular historical fact that the elegant soft hat of the Spaniard has remained the same from the earliest period to the present day, while among all other civilized nations a transformation in the style of that article has taken place. Comfort in the wear seems to have given place at all times to fancy and the demands of fashion.

Queen Elizabeth's patent grant to the hatters of

London is still recognized in England, and the 23d of November is the hatters' annual festival, that being St. Clement's day, the patron of the trade.

Preparation of Materials.

Previous to cutting the fur from the various skins, they must be moistened, straightened, and cleaned; the projecting long coarse hairs that are interspersed throughout the fur, removed either by pulling, clipping, or shearing; those of the rabbit, etc. being pulled, while those of the hare, etc. are clipped. To pull these superfluous hairs by the hand, the person sits with the skin laid over the knee, strapped down to the foot, and with a dull-edged knife in hand, the thumb being covered with a soft shield, the obnoxious guests are dextrously uprooted. If done by machinery, they are pulled out by being nipped between too revolving slender rollers. The skin is drawn over a sharp-edged board, which causes these hairs to project, and the rolls placed in the proper position and distance, frees the fur of its deteriorating associates with great facility, without disturbing the fur.

Furs intended for body-making undergo a process called carroting or secretage, which is an artificial method of increasing the felting quality of the fur, enabling the hatter to work at a kettle with clean pure water, dispensing with all acids and the like, and using boilers other than those of lead.

It is only of late years that carroting has been invented. It is a chemical operation or method of twisting or bending the natural straight-haired furs, and possesses also the property of raising or lifting the points of the scales which clothe the fibres of the fur, thereby facilitating the operation of felting;

while the fur in its original straight state could be used with satisfaction only as an outside flowing nap upon the hat.

The method pursued to accomplish this result is, to dissolve 32 parts of quicksilver in 500 parts of common aqua-fortis, and dilute the solution with one half or two-thirds of its bulk of water according to the strength of the acid. The skin having been laid upon a table with the hair uppermost, a stout brush, slightly moistened with the mercurial solution, is passed over the smooth surface of the hairs with strong pressure. This application must be repeated several times in succession, till every part of the fur is equally touched, and till about two-thirds of the length of the hairs are moistened, or a little more should they be rigid. In order to aid this impregnation, the skins are laid together in pairs with the hairy sides in contact, and put in this state into the stove-room, and exposed to a heat in proportion to the weakness of the mercurial solution. The drying should be rapidly effected, as otherwise the concentration of the nitrate of mercury will not produce its effect in causing the retraction and curling of the hairs.

No other acid or metallic solution but the above has been found to answer the desired purpose of the hat-maker, although sulphuric acid without the quicksilver has a limited effect when the skins are treated as those above described. For other purposes, such as that of the upholsterer, hair is curled by first boiling and then baking it in an oven; or it may be spun into ropes and baked, after which it is teased asunder.

Preparatory to cutting the fur from the pelt, the skins are dampened and flattened; they are thus made smooth and ready for the operation, which is performed by hand, with knives about two inches long by four wide, having a short upright handle. The skins are held upon a cutting-board, and the pelt kept moistened with water; a sheet of tin is laid upon the skin, pressed down by the left hand, whilst the knife in the right hand, being guided by the edge of the tin, is run rapidly forward and backward across the skin, gradually sliding the tin toward the tail; by this means the fur is gathered up, and kept in one fleece.

The pelts are appropriated to the manufacture of gilder's cement, or will make excellent glue. Machines in the form of revolving shears, similar to those used for dressing cloth, are employed for such skins as are uneven in the pelt, and which cut the pelt from the fur in slender shreds, being quite the reverse of the hand method, which cuts the fur from the pelt.

Stiffening and Water-Proofing Materials.

There is reason to suppose that when hats were first invented and long subsequently, the quantity of stuff or material weighed out for a single hat was of itself considered sufficient to stand unharmed the drenchings which it was likely to encounter.

However, such a hat in the warm season being unpleasant, a lighter body was proposed, to contain some stiffening substance as a substitute, and the attempt proved quite successful. A search was instituted for something suitable for the purpose that would harden the hat sufficiently, without increasing the weight, but rather diminish it.

In those times chemistry was comparatively unknown, and glue being at hand, our predecessors in the hatting trade commenced the stiffening of their hats with that material, which long continued the only article likely to succeed. Latterly, however, glue has become quite obsolete, having been entirely superseded by the various gums and resins, which, when properly prepared, enable the manufacturer to put into the market a much superior hat, and one more pleasant to wear, weighing 3 oz. which in former times would have weighed full half a pound.

The solubility of glue in water was its defect, and the ultimate cause of its rejection. Our spirited predecessors in the business, by a knowledge superior to that of their predecessors, coupled with a devoted spirit and unfailing resolution, after many vexatious trials but little known to our modern workers, succeeded in rendering a hat not only stout, light, and waterproof, but cheaper and more beautiful to look at, ventilated, and altogether pleasanter to wear.

Upon a retrospective view, and considering the total of these improvements, we may well excuse the many secrets and partialities existing in the trade, for before any new admixture of stiffening materials or method of applying them, whether before or after dyeing, &c., could be properly proved, many dozens of hats were under way. It required a length of time to enable a proper judgment of the experiment to be pronounced; thus, if unsuccessful, involving the character of the manufacturer as a tradesman, and his pecuniary affairs at the same time.

The result, however, was at last satisfactory, and now there are several methods of stiffening with a water-proof stiff, which possesses all the requisite qualifications.

There is no department in the hatting trade of more importance than that of stiffening, as the kind, quality, and quantity of the stiff must be regulated according to the country in which the hats are to be worn.

England, for instance, where there is so much moisture in the atmosphere, requires a much harder stiff than we do in America. American manufacturers finding that shellac possesses every requisite for both stiffening and waterproofing, now for their best hats use that gum only dissolved in alcohol.

> 20 lbs. orange shellac being dissolved with 5 gallons alcohol in a close vessel, *cold*,

attending carefully to stir it up repeatedly to keep it from lumping and sticking to the bottom. The vessel commonly is used in the form of a barrel or some sort of churn. When fully melted the stiff is ready for use by being thinned down to the desired consistency with additional alcohol and put into the hat with a stiff brush.

A cheaper, called alkali-stiff, and much used for inferior hats, is—

> 9 lbs. shellac, dissolved with 18 oz. of sal soda in 3 galls. water in a tin vessel.

The vessel with the water is set into another containing boiling water, and heated; the soda is introduced gradually, and is soon dissolved, and the lac is then put in and stirred occasionally for about an hour, by which time the lac will be dissolved. The whole is then left for an hour or two, when it may be taken out and set to cool. It is better if allowed to remain a few days after having been made. When used, it is reduced to the required strength with more water, a hydrometer being employed as a test.

The bodies are simply immersed in the liquor, and

passed between a pair of rollers one by one, thereby sweeping off the superfluous compound, but leaving them completely saturated. The hats with this stiffening must be immediately and rapidly dried in the stove.

This stiff is rendered the more popular by adding 3 oz. of common salt to the mixture before using it, as the salt neutralizes the soda, and the hats may be blocked immediately after being stiffened, thereby saving time and dispensing with the use of the stove.

The two following receipts are given as good and reliable English methods of stiffening hats:—

7 lbs. of orange shellac.
2 lbs. of gum sandarac.
4 ozs. gum mastic.
½ lb. of amber resin.
1 pint of solution of copal.
1 gallon of alcohol or of wood naphtha.

The lac, sandarac, mastic, and resin are dissolved in the spirit, and the solution of copal is added last.

This is called spirit proof, and like our own is put into the body with a stiff brush, and, being fully saturated, is set to dry.

A cheaper stiffening, also like our own called alkali or water stiffening, is—

7 lbs. of common black shellac.
1 lb. amber rosin.
4 ozs. gum *thus*.
4 ozs. gum mastic.
6 ozs. borax.
½ pint solution of copal.

The borax is first dissolved in 1 gallon of warm water.

This alkaline liquor is now put into a copper pan

heated by steam, or it may be set into another vessel containing boiling water, and the shellac, *thus*, and mastic added. This is allowed to boil for some time, more warm water being added occasionally, until it is of a proper consistence, which is known by a little practice. When the whole of the gums seem dissolved, half a pint of wood naphtha must be introduced, and also the solution of copal, the liquor should be passed through a fine sieve, when it will be perfectly clear and ready for use. This stiffening is used hot with the following preparations.

The hat bodies, before they are stiffened, should be steeped in a weak solution of soda, to destroy any acid that may have been left in them. If sulphuric acid has been used in the making of the bodies, after they have been steeped in the alkaline solution they must be perfectly dried in the stove before the stiffening is applied.

When stiffened and stoved, they should be steeped all night in water to which a small quantity of sulphuric acid has been added. This sets the stiffening in the hat body and finishes the process.

If the proof is required cheaper, more shellac and rosin may be introduced.

The Blowing Maohino.

In the manufacture of the finest kinds of fur hats, namely, those with a flowing nap, the stuffs of which they are made must be thoroughly refined.

The clipping and pulling operations, to which the skins were subjected previous to cutting off the fur, never free the fur entirely of the coarse hairs that are intermixed with the finer; and to separate the coarse from the fine, the fur, as it came off the skin, is placed under the action of the blowing machine, which consists of a long, close, narrow, wooden box, divided into a number of apartments, the divisions between each of them having an open space at the top or bottom, so that a blast of wind can be propelled through the whole length of the trunk. The fur is put into one of these receptacles at one end, where it is teased and tossed by revolving brushes set in the bottoms of several of them, and a revolving fan is placed at the head. The whole being set in motion by some first power, the blast of wind from the fan seizes the loose thrown up fur that is tossed by the revolving breakers and brushes, and the stream of flying fur is transmitted from division to division, along the whole length of the wooden box. In this operation, the fur is graded as it is blown along and deposited gradually in the respective places, lodging in the most regular order, from the one end of the wooden trunk to the other, the dust and dirt falling down below, the heavier portion of fur not being blown to the same distance as that of the finer, which reaches the farther end, where the finest of all is received entirely refined of its impurities. But the cutting and blowing of fur are both independent and distinct branches of business, although relatively connected with that of hatting, and the various grades of fur are bought by the hatters from the professional hat furriers or their agents.

The Manufacture of Hats.

Before commencing a detail of the processes of the trade, it will be necessary to bear in mind that hatting is universally divided into two great divisions, viz., the making, and the finishing departments, each of which as a matter of course has its subdivisions.

With the exception of encyclopædias which give detached and very abridged descriptions of felt making and of hatting generally, there has been no specific account published in either pamphlet or book-form, so far as the writer is aware, of the manner in which felt hats are made, or of the principle of felting by which they are produced.

This is considered by the writer a sufficient inducement to illustrate to the best of his ability a principle entirely belonging to natural history, viz., the natural scaly clothing that is upon all hair, and hitherto but little known, and upon which several important branches of business depend. Indeed, it seems almost absurd to think that a hair, puny as it is in itself, bears upon its sides a something of such importance, so very minute as to require the utmost attention with the aid of the best microscopes to be seen at all, and yet upon that something is based the art of felting and of course of hat-making, besides several branches of other trades, some of which have already been mentioned.

Hat-making was long considered a business to which machinery never could be applied, but the inventions of man have at last dispelled this illusion, and machinery is now employed in several of the most important departments of the trade.

The reason why this idea obtained such general credence was, first, on account of the close attention requisite, while the hat is under the operation of sizing.

Second, the known impossibility of napping or ruffing a hat by any means with machinery, also, the acknowledged failures of several attempts to substitute carding for that of bowing, and various futile attempts with the irons in the finishing department.

The innovations of machinery, however, have now obtained a sure footing in all large factories, and some of them will come under observation in their proper places.

In the mean time we shall confine our observations to the old system, which still prevails in most small factories and all small towns.

Our honest forefathers, the manufacturers in former times, would insist upon making hats to wear not for a season, as with us, but for many years, being afraid of damaging the trade to do otherwise, but now a hat for city wear, of scarcely three ounces weight, and lasting two or it may be three months, is quite a common thing.

The usual quantity of stuff given out for a regular felt hat, modified of course to a very great extent by the market, we shall suppose to be three ounces of fur. It may or may not be a mixture of different kinds and qualities of stuff previously prepared by carroting, and may or may not be refined by the winnowing machine, which separates the different qualities of fur. These three ounces, however, are sometimes increased by unprincipled men to four and a quarter or four and a half ounces, by the addition of other and cheaper ingredients, which are all laid upon a platform of boards about five feet square, called a hurdle, over which a large bow of about six feet long strung with cat-gut, Fig. 1, is suspended. This bow is held by the left hand of the hatter, and with the right he holds a small piece of wood with a head or knot upon it, Fig. 2, with which he tugs the string of the bow and makes it vibrate upon the stuff, and into it, with great dexterity and with the nicest judgment. This operation has always been considered a beautiful sight

Fig. 1.

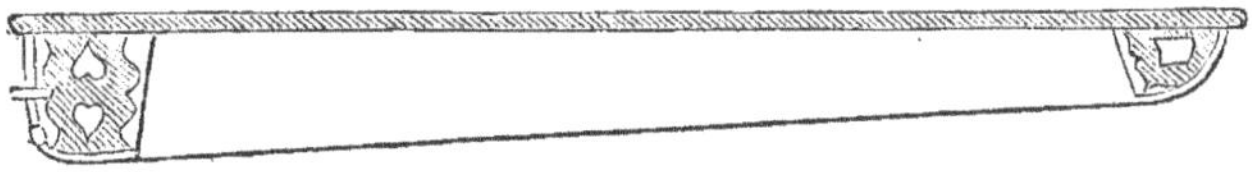

Hat Bow.

to a stranger, as the performer goes on plucking the string, and the string playing upon the top of the fur,

Fig. 2.

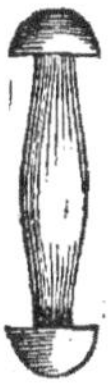

Bow-Pin.

which lies upon the left hand side of the platform. The fur touched by the string is made to fly from one side of the boards to the other with the greatest regularity. So nicely is this bowing performed, the stuff flying from the bow-string hair by hair, and flake by flake, that a hat in this loose state may measure several inches in thickness.

In this operation, the different materials are tossed about to-and-fro repeatedly, and mixed with a much greater regularity and change of position of the various filaments than if drawn by carding machinery. One half of the intended hat, called a bat, is bowed at a time, and both in nearly a triangular shape, which being gathered up, and pressed with a flat square piece of wicker-work, Fig. 3, and afterwards with a smooth skin or cloth, is pressed and gently rubbed with the hands backward and forward so as to create a friction on the surface fibres, thereby interlacing the outside filaments, by which means the simply safe-lifting of

Fig. 3.

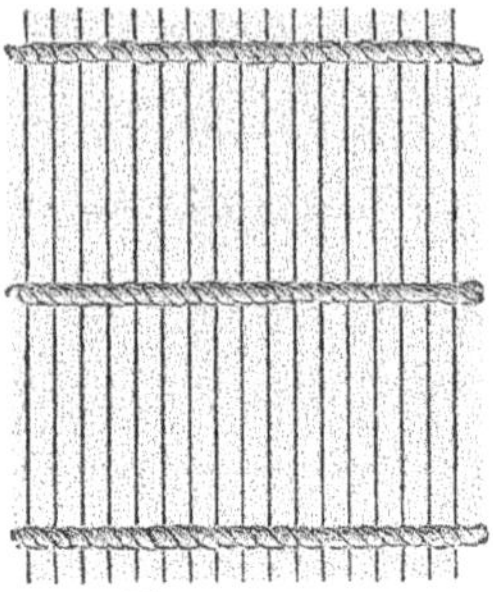

Hat Basket.

these two half-solidified portions of the future hat is secured. The one-half being laid upon the other, with a triangular piece of paper or cloth between, they are joined together by overlaping two of the three sides, thereby giving to the intended hat the form and figure of a hollow cone or great bag, but so tender that none but an experienced hatter could handle it.

This operation of bowing is the same, with but little variation, whether it be for coarse or fine hats.

If wholly of wool, they are now swaddled carefully in an outer cloth, and sprinkled with water, and laid upon a warm plate of metal which sends the steam up through the hat which is to be pressed, and slightly rubbed, sprinkled again, and turned over. Continuing the pressing and rubbing, and by repeating these operations for some time, the motions are transmitted to all the inclosed fibres of wool with an irritating feeling, as it were, exciting their propensity for travelling, till the outer hairs, in their motions, warp themselves with each other and the surface appears skin-like and becomes smooth.

During these actions, the hat inside of the cloth must be several times changed in position and kept in proper form, when its swaddling envelope and the paper within which kept the inside open and free may be removed. These operations concluded, the tender hat must now be subjected to a much more laborious operation, where, properly speaking, the grand practical art of felting takes place, where thousands of thousands of filaments are all in active though slow motion, all travelling on their own individual course, independent of, and at the same time dependent upon, each other for their mutual support, being carefully guided collectively, by the hatter's good judgment.

This stage of the operation is a wet one requiring an open boiler surrounded by planks, which slope towards the centre, called a battery, Fig. 4, suitable for six or eight men to work at. Each man is provided with a rolling-pin, cloths, brushes, etc. The soft and tender hat is laid upon one of these planks or benches,

Fig. 4.

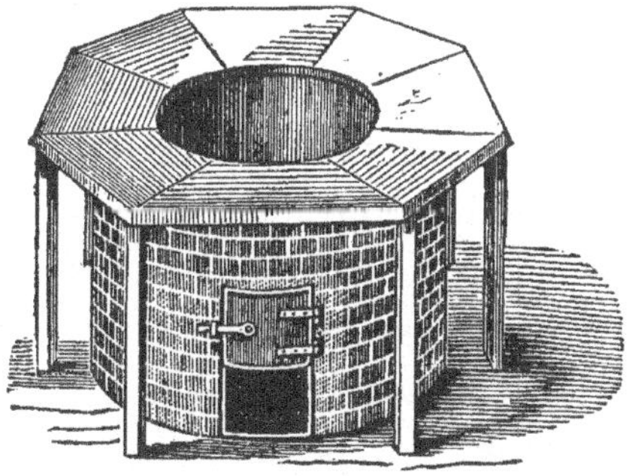

Battery for Sizing Hats.

wrapped in a damp cloth, and carefully wetted, squeezed, folded, rolled and unrolled, keeping it constantly moistened by dipping it in the hot water of the boiler, folding and unfolding with every variety of crossings, rolling it as a scroll, pressing, shaking, dipping and rolling it again and again, the hatter all the while bending over his work in front of the almost boiling caldron, and surrounded by steam. He labors hard, ever changing the position of the hat under his hands, so as to make it an evenly felted and perfect piece of work, which these oft-repeated motions ultimately accomplish.

This is the grand felting operation; the cause of which was so long considered a mystery, and now ascertained to result from the peculiar natural construction of the animal fibre, as already explained.

In this planking or sizing of the hat, sometimes with half a dozen under hands at the same time, the enveloping cloth is soon thrown aside as the hat grows in solidity. The hands of the hatter are defended from the scalding water by thick leather shields upon the palms, and as the hat approaches its proper size, it is scalded and belabored with determined importunity, coiled, rolled, pressed, and pinned, backward and forward till the size of the hat is reduced to nearly half of its original dimensions, and the tension of the several fibres becomes so great that the hat will felt no farther. At this stage it is impossible for it to be torn asunder, and is still in its original form of a hollow cone.

Such is the making department of the trade, the felting process, where a firm piece of cloth (for such is the body of a hat) is manufactured from loose wool or fur, independent of either spinning or weaving.

We have now explained the making of the bare

body, as it is called, of a plain hat, in as concise a manner as the subject will permit.

There are yet a variety of qualities and kinds of hats requiring a variation more or less in the manipulation of the article, so as to suit a fanciful and fastidious people. For instance, the quantity and quality of fur, or an entire change of materials, produce quite a different appearance both in the look, the wear, and the price of the hat, while the form of the cone must be changed to admit of a high or low crown, or of a broad or narrow brim, &c. &c.

All FELT hats, of whatever texture, nature, or name, must have undergone the above described operations, and many have to go back a second time to the plank kettle, and there undergo an additional teasing and ducking in the scalding water. For instance, all those destined to receive a coat of fur upon the outside finer than that of which the body is made, and constituting the flowing nap of the hat, which is merely a kind of veneering or outside plating, which will shorly be described.

A very good hat is made having a flowing nap that is raised directly from the body itself. Thus when the body of such a hat as has been described is about half wrought up at the kettle, it undergoes in another department the operation of shaving, by which means the projecting coarse hairs are all cut off, after which, on being returned to the kettle, the hatter, with his stiff brush, card, and comb, raises a nap upon the half solidified body, which is constantly improved as he continues to manipulate with the brush. The hat is, at the same time, reduced in its dimensions by the operation of felting until at the conclusion when it appears of the desired size, fully felted, and adorned on the outside with its rough and flowing nap, which otherwise would have been smooth and clothlike. This is called the brush hat.

Shaving.

In the process of fur felting there is a constant tendency for the strong straight hairs of the body to work to the outside, so that whether the hat is designed to receive a BARE finish afterwards, or to get a plated cover of beaver for a nap, those bodies must all undergo the operation of shaving. A workman sits in another apartment with one of them, when dry, spread over his knee, and with a long bladed sharp knife in hand, sweeps rapidly over the surface, cutting off and depriving it of those deteriorating superfluous intruders, after which the hats are forwarded to the stiffening department.

Stiffening Process.

The bodies of the hats now made, dried and shaved, and the spirit water-proofing already prepared, being thinned, or reduced to the proper consistence, the hat is laid upon a flat sloping board, and the stiffening is put into it with a stout brush, and soaked to that degree of saturation known only by experience, the brims receiving a double portion for extra stoutness, and are then set aside to dry.

The alkali or inferior kind of stiffening, when used, is likewise diluted, and applied by immersing the body fully into the prepared ingredients already described, and either wrung out with the hands, or passed a couple of times between a pair of rollers set at a proper width, which determines the quantity of proofing absorbed by the hat.

It should be observed, regarding this stiffening of hats, that it is simply a varnishing of the several fibres of the fur of which the hat is made, each hair individually has got a coat of waterproofing varnish, for when dry it will be found that the interstices between each and every fibre are quite open and free, and therefore susceptible of ventilation; thus differing entirely from what would have been the case had it been stiffened with any kind of paste.

Ruffing or Napping.

Very little of this is done at present in the United States. After the bare body of the hat is stiffened, if a flowing nap of beaver, otter, neutra, or other fine fur is desired, finer than that of which the body is made, half an ounce more or less of the superior uncarroted stuff is weighed out, sufficient to cover the whole outside surface of the hat. The hatter lays this precious morsel with perhaps one-eighth ounce of cotton on the hurdle, under the bow, as he did with the stuff for the body, and with a similar but lighter instrument, these two stuffs are completely mixed and spread upon the boards, as evenly as his experienced hands can do it; the cotton being used merely to enable him to handle the fur, which otherwise would be so thinly spread, and so attenuated of itself, as to endanger the simple act of lifting it. This mixture of fur and cotton is next spread upon the wet bare body of the hat as it lies upon the plank at the kettle, a little water is sprinkled over it and beat down with a brush. The hat with this surface covering is wrapped very carefully in a piece of cloth or coarse hair-cloth, and operated on very lightly, and nearly in the same manner as when felting the body. The object to be attained is to get the fibres of the fine fur to penetrate the body, and take root as it were therein—great care and watchfulness being demanded of the workman at every motion of his hands, in this manner of working. The points of the fibres of the beaver fur penetrate the body of the hat, and having once got a footing, it constantly advances, as the active careful rolling, folding and unfolding, shaking and tossing go on, until the fur has separated itself from the cotton; by its boring, having obtained a firm lodgment in the solid felt of the hat body root end foremost. The cotton with which it was mixed is left behind loose and useless, for want of the little rough scaly property that the other possessed. An inexperienced workman in thus ruffing a hat is liable to continue his work too long, until the beaver napping has burrowed quite through to the inside of the hat, where it is lost.*

In the various operations of the hatter with hot water, whether in body-making, napping, or dyeing, &c., the water should not be allowed to boil, for independent of the damage to some kinds of stiffening, as hair contains a large portion of gelatine in its substance (to which alone it owes its suppleness and toughness), this gelatine will be separated from the hair. This is particularly the case with napped hats, for when thus treated the fibre becomes much more

* Hatters' kettles for FUR hatting are made of copper, as they take less firing than those that are made of lead. But lead must be the metal if sulphuric acid, no matter in how small quantity, is used in the water. It is generally resorted to by the men in sizing WOOL hats, as it facilitates the felting operation. This acid (vitriol), having no affinity for lead, does not affect the kettle, while it would soon eat the one of copper through. Care, however, must be taken that no stone be let fall into the water of the lead kettle, for a hole will soon result from such an accident.

brittle than before, and the nap soon breaks off round the square.

Fur hats having a flowing nap are sometimes clipped very short with revolving shears similar to those used in dressing cloth, and which is done previous to blocking or dyeing.

Fig. 5.

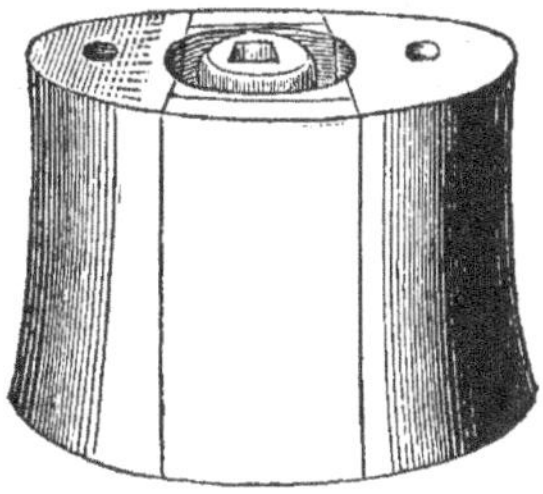

Hat Block.

Blocking.

Previous to dyeing, all hats must be blocked, using such blocks as approach the intended shape of the hat, and as soon as possible after the making department is concluded. It is a laborious operation, though simple, as the nature of felt allows it to be stretched to a great extent in any direction when it is wet and hot.

In the act of blocking, the conical form of the hat is lost for the first time. The hat is now immersed in the *boiling water* of the kettle, and while wet and hot the tip is stretched wide, and the whole thing simply drawn down over the block, a tight cord is run down to where the band is to be and the brim flattened out.

Dyeing.

The next operation is that of dyeing or coloring, and if convenient, and the hats fine, each hat should be upon its respective block when in the color kettle.

great care being observed to keep the square from abrasion, as the least rub may deprive a napped hat of its fur at that exposed and important part. Most generally, however, the hats are colored without a block, the blocking being performed as soon after the dyeing and washing as possible in *boiling* water.

The ordinary ingredients for black are, for 12 dozen,

144 lbs. of logwood, *chipped*, or its value in extract.

12 lbs. of green sulphate of iron or copperas.

7½ lbs. of French verdigris.

The kettle should never boil nor exceed 190 degrees, and during the operation the hats must be repeatedly taken out and exposed to the action of the oxygen of the air, so as to strike a deeper color, and during the necessary exposure to these airings, the time is improved by having two suits of hats going on at the same time.

From six to twelve hours are required to complete the operation. The shorter the time the hats are in the dye, compatible with the deepness of the color, the better will be the goods, as boiling extracts the gelatine of the hair and makes the nap brittle, which is seen by comparing dyed articles with those that are of a native color.

Pumicing or Pouncing.

Pouncing is a term for rubbing down the outside of a hat with a piece of pumice stone, sand paper, or emery paper, whereby the hat is made entirely bare, smooth, and fine, resembling a piece of very fine cloth. These are generally called cassimere hats. This operation is usually performed after dyeing, and previous to finishing. Some makers, however, prefer to singe the hats instead of pouncing, but such hats never feel so fine as the others, as the singing of any hair invariably produces a hard crisp burnt knob upon the end.

Finishing.*

When a hat arrives at that state of forwardness ready for finishing, it is a very unsightly object to any person but a hatter. Most of its processes have been wet ones, but now it is to assume a genteel and prepossessing appearance, under the artistic appliances of brushes, cloths, hot irons, and labored exercise. If a plain soft hat, it is pulled over such a

Fig. 6.

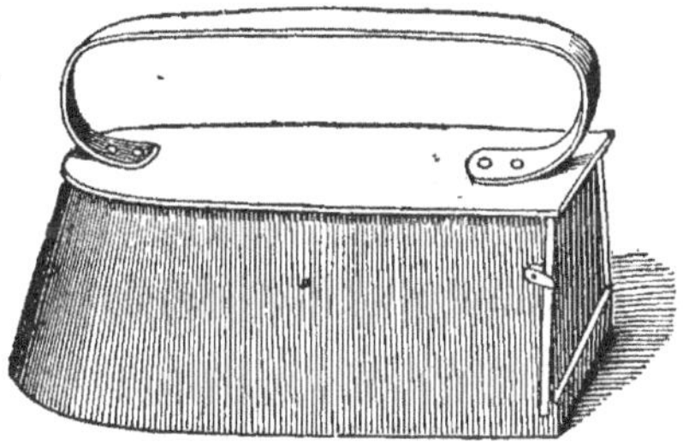

Hat Iron.

block as is required, a cord is run round the hat to keep it tight upon the block; the tip and brim are then flattened with the hot iron, wet sponge, brushes, and hair-cloth cushion or velure, several wettings being necessary in finishing.

The brim is next cut to the required width, and the cord run down to the depth of the block. The

* As every hat must be finished upon a block of some particular form, upon which the hat assumes the exact counterpart, it becomes necessary with those having broad tops, that the block be in five separate pieces, so as to get them out or into the hat, the centre piece acting as a wedge to the whole.

side-crown is now to be finished, along with the tip and upper and under sides of the brim, the hatter exercising his best judgment. The block is then withdrawn, the brim curled and set, and the finished hat sent off to the trimmer to get lined and bound; it is then tipped off and packed for market.

The finishing of this kind of hat is a simple operation when compared to that of a napped hat; requiring only the assuming of the proper shape and form, the solidifying of the body, and giving it such a lustre and finish as the quality of the material will allow.

The stiff cassimere hat being less flexible, is subjected to hot steam preparatory to blocking, whereby it is made soft and pliable. When in this state it is drawn down over the block, and the block withdrawn, to insert a prepared disk of pasteboard into the crown for strength, after which it is finished much in the same way as that already described, but with the difference, that a cloth must always intervene between the hot iron and the hat when finishing.

The finishing of a napped hat, whether it be brush or beaver, is a very different process from that for either of those just described, requiring the nicest attention and patient perseverance by the best workmen. The hats are given out by the half dozen, which are sorted for the different sizes and steamed one by one; the hot steam softens the stiffening, and when pliable, the hat is drawn down over the respective finishing blocks, the nap of each hat straightened with a wet brush, and a half finish given to it with the water, brush, and bare hot iron. The block is then withdrawn and the hat given to be shaved with a razor. This seems a singular operation;

but a few passes with that instrument over the hat effectually cut off all those projecting coarse hairs that have eluded all previous attempts at removal, and without in the smallest degree endangering the finer fur of the nap. The hat is now returned to the finisher to complete the process.

These coarse hairs, when left in the hat depreciating very materially its value, were formerly plucked out by hand with a pair of pickers, hair by hair, often to the injury of the hat. The advantage of the razor will be obvious to all.

A pasteboard disk, well spread with dissolved shellac, is now inserted into the tip, and the block reset. The workman with his hot iron, wet and dry brushes, &c., lays down the nap in its proper direction, and the hat by continuous labor becomes solidified and more elastic, the tip is rendered stout by the adhesion of the prepared inside disk; and by the repeated wettings, and careful ironings and brushings, all the ripply appearance of the fur is destroyed, and the whole surface becomes smooth and shining.

The crown being finished is then papered up, and the same operations that were bestowed upon the crown are now to be repeated on the brim, both on the upper and under side, which having been accomplished, a gauge is applied and the brim cut to the required width ready for the trimming.

There is a beautiful invention for preserving the form of all hats having flat or soft supple brims by means of a flattened wire, upon which two small twists are made, and when joined as a hoop, the proper concave is produced. This hoop is attached to the outer edge of the brim, and covered with the binding, and thus the unsightly slouch that often deformed, particularly the soft brimmed hat, is permanently prevented, and the graceful curve completely secured.

Silk Hatting.

The art of silk hatting is comparatively of modern invention, consisting simply of a cover of silk plush over a body of some other material. As much sleight of hand is required in this department, it naturally follows that a good workman is a valuable and appreciated artisan.

The bodies used for this kind of hat have been so various, that a full, or even succinct, description of them would be quite superfluous. Wool and fur bodies, straw and leghorn, cork, whalebone and muslin, &c., even stretchers similar to umbrellas without a body at all have been adopted, and all of them have had their day. At present, however, the trade seems to have settled down to the two kinds—fur and muslin.

The fur body of a silk hat, called a shell previous to coming into the hands of the silk finisher, is made much in the same manner as that of a plain soft hat, by felting and sizing it down to the proper dimensions in the plank kettle. It is quite light and thin, and when blocked or otherwise, and dried, is then ready for *stiffening* by the finisher.

The different substances for this purpose, and the various methods of doing it, have been as numerous as the varieties of bodies that have been adopted. The whole of them, however, now have been abandoned for shellac.

The most simple and the best stiffening for any hat is shellac dissolved in alcohol, and thinned down to a proper consistence. A cheaper, however, and at the same time good stiffening, is the ammonia stiff already described. Either of these is applied in a like manner and with the like operations. The soft body or shell, as it is often called, is immersed in the liquid in a basin, then wrung out and pulled upon a block, the brim being flattened, a brush is dipped into another ves el containing a thicker lac, and applied to the square and brim for extra strength; after this the block is withdrawn, and the body set to dry.

These felted bodies or shells, as they are called, when dry are steamed generally over the hatter's hot iron, and pulled when warm and soft over the finishing block. A cord is then run tight round the shell, and the block withdrawn; the prepared pasteboard tip is inserted into the crown, and the block reset; after which the body receives a regular hot ironing all over. In this operation the inserted tip adheres to the felt, and the whole body assumes the exact counterpart of the block, both crown and brim. The rough hairs are now to be removed by sand or emery paper and the block withdrawn. The body next receives a coat of the best size, and when dry two coats of seed-lac, or copal varnish which finishes the making of this kind of body.

Those bodies that are made of muslin, when first invented, were called gossamer, from their extreme lightness, and though they have increased in weight, they still retain the name of gossamer hats.

In preparing for the body, a few yards of muslin are extended upon a frame, and saturated with liquefied shellac, or water stiff, which when dry is cut bias into strips for sides, tips, and brims. One side of these side and tip pieces of muslin is overlaid with the silk intended for the inside lining of the hat, and pressed to adhesion; or this may be done while in the web before being cut into strips. The block being set upon the bottom-board, one of these extra prepared sides is wound tight round the side-crown of it, and the two ends stuck together by overlapping. A piece of the prepared tips is next laid on, and made to adhere to the side-crown.

The brim consists of three thicknesses of stout muslin of a circular form, each with a hole in the centre, all of which are slipped over the crown down to their place of destination with a quarter of an inch of the edge rising up on the side. A second side-crown and another tip are now applied, covering the others, and the whole of these cemented together with the hot iron, the shellac with which they were stiffened acting as a cement. After receiving a coat of size and one of varnish, this body will be ready, like the other FUR body, for the finisher. In preparing these bodies, cover the block with a soft shell.

Before commencing the finishing, however, we will describe the sewing of the silk plush cover, which is quite a nice and particular piece of work. The strip of plush for the side-crown is cut from the web bias and of a width the depth of the intended hat; the tip piece which is to mate this side-crown is of course circular, and a quarter of an inch larger all round than the tip of the hat. These two pieces are to be sewed together by hand face to face, the edges being folded back, and the plush put well through to the proper side with the needle as the sewers proceed, so that the seam when the hat is finished may not appear bare for want of plush.

In finishing, whether the hat body be of fur or

gossamer, the first thing is the putting on of the under brim, which we shall suppose to be plush, satin, or merino. A strip is cut from the web or piece at about an angle of forty-five degrees, and having the length reduced to suit the size of the hat; the two ends are then sewed together, and having been laid on the hat, one of the edges is made fast to the edge of the brim with the iron, all round, and smoothly laid down, the bias allowing this to be done by stretching. It is next to be steamed with a damp cloth under the hot iron and the inner edge stuck inside of the hat with the nose of the iron.

The upper brim is next in order. A strip of silk plush the requisite width is run on, slightly, in much the same manner as with the under brim, but dispensing with both the cloth-steaming and often with the sewing. The one end of this upper brim being cut with the scissors and the other with the knife, a good invisible seam may be made.

The brims being now on, the tip of the hat is wetted inside, and the block put in. The silk plush cover, having been previously spread with gum tragacanth, about where the side seam is likely to be, and now dry, is carefully drawn over the crown and fitted to the hat; the two ends of the cover being folded back and marked for the seam. The cover is then removed, the plush brushed back at the folding, and the cloth cut for the seam with a pair of sharp scissors; the top of the seam is cleaned or dressed off and the cover replaced on the hat body. The tip and side crown are now to be stuck with the hot iron to the body with particular care, so as to make a good joining at the seam, and not to draw through the varnish. The making of a good seam is the test of a good workman.

The dressing and polishing of the hat now commence; and while it remains upon the block, this is done by means of brushes, wettings, ironings, etc., once, twice, or three times in succession, after which it is fixed on the veluring machine where it is revolved rapidly, for the purpose of freeing the nap of all impurities by means of the hair-cloth velures that are applied.

The hat is next taken back to the bench, where it receives its final dry-ironing, veluring, etc., and the crown is papered up.

The brim is yet to be finished, which is done by hand, with the brushes, sponge, iron, etc., and made to shine like the crown; after this it is given to the trimmer to be trimmed and bound, when it comes back to be curled and properly shaped in the brim, suiting the taste of the wearer. The workman who gives the hat its final touches makes use of a number of tools, which, though of seemingly trifling appearance, are nevertheless necessary for his department, which requires a refined taste.

Forming Machines.

Such is hat-making, but we cannot conclude without remarking that there have been many patents granted in this and other countries for improvements in hatting, that we cannot notice. Nevertheless there are two, of decided merit, claiming attention, as having entirely revolutionized one-half of the making department, and which may be modified and extended to answer many purposes, in addition to that of hat-making.

Fig. 7.

Forming Machine for Fur Hat Bodies.

The first and most ingenious is called the pneumatic process of forming the bodies, hence in all large cities the bowing operation is not employed. It is as follows: A cone of sheet copper punched full of small holes, and set upright, revolves slowly upon its axis; beneath this or attached to it an exhausting fan is placed, causing by its rotation a current of air to draw through the holes from the outside. A trunk or box with an opening facing against this revolving cone, discharges the fur which is fed into it at the other end by a feeding apron, in quantity just sufficient for one hat-body. It is drawn into this trunk between two rollers that are covered with leather or felt, and immediately seized by a cylinder revolving about four hundred times in a minute, furnished with a number of stiff brushes. This generates a current of air which scatters the fur and blows it out of the mouth of the trunk, where floating in the air it is speedily drawn upon the perforated cone, and evenly spread over the top and sides of the same, in quantity enough for one hat-body in so many revolutions. The discharging trunk is so adjusted that any desired quantity of fur can be deposited on any particular portion of the cone. When the cone has got the fur for one hat-body, the workman wraps over it a wet cloth and slips a metallic cover over the whole, which he removes into a tank of hot water. A new cone is immediately set in its place to receive another coating of fur. The hot water into which it has been dipped tends to make the mat more tenacious, which is next slipped off the cone, taken to a table, gently worked by hand-rolling in a piece of blanket, squeezed and pressed, and folded into a convenient shape and sent to the regular hatter to be felted at the ordinary plank kettle.

The cost of hat-bodies is reduced, it is computed, by this process as five or six to one of the old bowing system, and the rapidity of production is as thirty to one.

It will not have escaped observation that this ingenious piece of machinery is applicable only for fur, the filaments of which are short and less inclined to tangle than those of wool, but another and no less useful piece of mechanism has been invented for forming the bodies of wool hats, and like the other has entirely superseded the use of the bow in all large factories where wool hats are made. It consists of a modified common carding machine, the sliver from which is conducted to a set of double conical blocks that are placed base to base, and which slowly revolve upon their axes in front of the carding machine, and the sliver is received and wound upon these combined blocks to the required thickness,

Fig. 8.

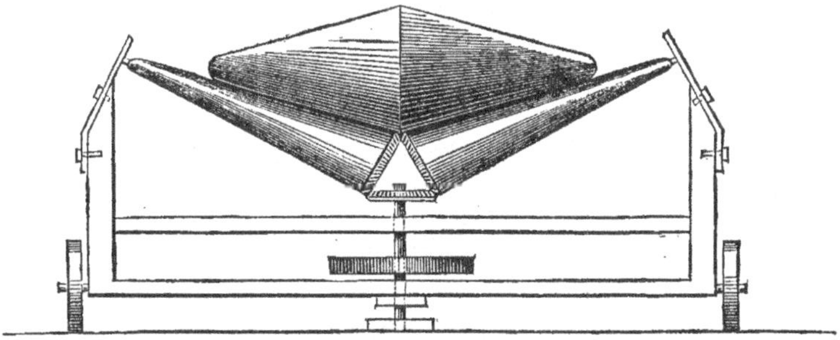

Forming Machine for Wool Hat Bodies.

sufficient for one hat, both blocks being covered at the same time. This machine which carries the blocks has a horizontal vibratory motion, or swaying backwards and forwards, that enables the sliver to be wound in a systematic manner on the cones, with a varied thickness of material for brims and crowns, and causing also the fibres of the wool to lie in a diagonal position, as layer upon layer covers the blocks. The bodies of the two hats, each of a conical figure, are thus made over the surface of a double cone which are separated by cutting them along their middle or base line, and slipping them off at the end. They are now ready to be wetted, shrunk, and felted in the usual manner by the regular hatter.

Shoes and Gaiters of Felt.

We will here describe the making of felted gaiters and shoes, which is similar to the art of hatting. There may be other and better methods, as the expansive stretching nature of felt may admit of other modes.

The wearer of these gaiters may walk upon the slippery pavement with comfort and full confidence, and if furnished with a leather or rubber under-sole, they are a neat, easy, comfortable cover for the feet.

A given quantity of wool calculated for one pair of shoes is weighed out, which is divided into four equal portions, two of them for each shoe. One at a time is laid upon the hurdle, and with the proper bow it is bowed as if for a hat, and disposed of in exactly an equilateral triangle, which being gathered together with the basket, is pressed, and temporarily solidified, laid aside, and the other portion treated in the same manner. A piece of coarse brown paper is now folded into a triangular shape, a little smaller than the bats just bowed; all the three edges are to be folded together with the paper inclosed. The use of the inclosed paper is to prevent the inner surfaces from felting together, and to keep the inside open. The intended shoe is next lapped in a sheet of cloth, and hardened at the hot basin (the basin is a disk of solid iron with a fire beneath). Water sprinkled on the sheet when turned upon the basin, sends steam all through the mass, and when rubbed slightly by hand, friction is communicated to the surface fibres, which in a short time become smooth, when the position of the triangular wool should be changed and the rubbing continned. A few crossings and rubbings give it a consistence sufficient for handling at the plank kettle, where we shall suppose both shoes to have arrived.

The felting operation at the kettle is performed in quite the same manner as that of a hat, by pressing, rolling, folding, and unfolding, &c., with its dippings into the hot water, until the material has assumed a hardness and solidity quite astonishing to the casual observer.

This operation finished, the shoe still in the triangular shape, one corner is now to be cut off to make an opening, and the confined paper taken out, which is quite a soggy spongy lump of pulp. The mate to this shoe having been advanced to the same forwardness, they are to be pulled upon their respective lasts and dried, and perhaps dyed, after which they are pounced, and finally trimmed.

Printers' Sheets.

The making of *sheet* felt for calico and other printers is a business that fell into the hands of the hatters at the introduction of the waterproofing of hats, as previous to that time the thick stout old hats of former times were quite sufficiently thick for the fittings of their blocks, so that when no more of them could be gotten, recourse was had to the new article, although it should be at a slight sacrifice.

Felt is employed in this business because of the facility with which it lifts and carries from the color sieve, the colors that are to be applied to the cloth. Wood and Copper blocks or rollers require two different thicknesses of felt, and though various qualities are made, a solid body and clear smooth surface and edge when cut and pounced by the block-cutter, are absolutely necessary, as otherwise, a ragged edge to the printed figures on the cloth will be the result.

The following makes a very good article:—

> 7 ozs. best backs of coney wool, and
> 6 ozs. of Saxony lamb's wool.

The coney is first well broken over with a light bow, upon the hurdle, and then by means of the heavier wool bow, the well-carded Saxony is intimately mixed with it. This thoroughly accomplished, the whole is to be divided into two portions; the one a little heavier than the other, which is laid upon the hurdle, and with the same wool bow, strung with stouter cat-gut, the hatter disposes of the mixture in a perfectly even flat form, of an oblong square, which when gathered by the hatter's basket, measures 18 inches wide by 3 feet long. A cloth is then spread over it, and the whole turned upside down; the sides and ends of the cloth are lapped over, so that this bat as it is called is completely enveloped. A stiff skin is now thrown over it, and pressed and rubbed for some time in an even manner, to reduce its thickness. The skin having been removed, the sheet with its bat is rolled and pressed still more, then laid aside while the other half undergoes exactly the same operation, but is made three inches shorter in length.

These two sheets, which are destined to form but one, are connected thus. The shorter is first folded over upon itself, and the two ends joined by overlapping with a proper inlayer of paper; then the larger bat is laid upon this one, and the whole turned upside down, so that the joinings of the two bats will be upon opposite sides of the sheet of felt. After these joinings are carefully made, the would-be sheet appears exactly like a lady's muff, and is again to be enveloped in the cloth, for the hardening process, at the hot basin, where it receives a partial steaming, rubbing, re-folding, &c., till finally it is carried to the plank kettle, where the severest labor must be applied; the object being to condense the materials of which it is made to the utmost degree of tension. It is then cut open, dried, and receives an application of a weak solution of size; when again dry it is well pounced with pumice stone, and the edges cut straight, which finishes a first class printers' sheet of felt, the size being 33 or 33½ inches long by 12 inches wide. Sheets for *copper* blocks or rollers require a thickness of a quarter of an inch, and those for *wood* three-sixteenths of an inch. Some prefer a sheet altogether of Saxony *wool.*

Cloth Hats.

After the introduction of gutta-percha into the arts, and the manufacture of it into thin sheeting, a new kind of hat was introduced, made of gutta-percha cloth, and from the variety of shades, &c. seemed for some time to supersede the soft low-crowned FELT article. But the cupidity of some of the manufacturers destroyed the business almost entirely when in its infancy, some say purposely, by making them so very inferior and at the same time so perfect a counterfeit, that the really good and perfectly made hat became universally distrusted, and hence the result.

We shall refrain from all notice of the methods employed tending to this deterioration of the new article, and merely describe the making of the honest, sound, and valid hat, the revival or resuscitation of which is well worthy of consideration.

A dry, thin, and soft fur or wool body is to be drawn upon the proper block, generally 3 or 4 inches deep with either a square or round crown, and the brim spread out upon the bench or bottom board. A circular piece of gutta-percha gum the size of the intended brim, having its centre cut out, is to be slipped over the crown down on to the felt brim; a similar piece of good cloth is likewise slipped over in the same manner to cover the gum, and now the extreme outer edges of the felt and cloth are to be carefully cemented together by means of the gum, by passing round a hot iron. The usual stirrup or bridle is then thrown over the hat, girding the inner edge of the cloth to the block, and stuck with the heel of the iron. This partially stuck brim is finally overlaid with a wet loose brim-cloth and properly ironed, the heat of the steam from the damp cloth softens the gutta-percha gum and effects the adhesion of the cloth to the fur body. About half an inch of cloth will project up on the side crown, which is also made to adhere to the felt body by the heated iron.

The block is now to be withdrawn, and the hat turned inside out, which reverses this would-be upper brim to the under side. The hat is next to be reblocked, a repetition of the gum and cloth is to be applied to this side of the brim exactly as with the other, and then succeeds the covering of the crown, which is to be wholly laid over first with the gutta-percha and then with the previously prepared cloth cover as a crown piece, these being held tight by means of the blocking-cord. The whole crown, both tip and sides, is to be cemented and finished, never omitting the wet finishing-cloth between the hat and the hot iron, and the hat is now complete and ready for lining and trimming.

The above makes a good soft pliable cloth hat. But if a stiffer and firmer hat is wanted of the same material, the felt body is to be put through the process of the alkaline bath, similar to that of fur hats, and when dry, proceeded with as above.

Another method of making these cloth hats is to dispense with the fur body entirely, the block being covered with two thicknesses of cloth and having a ply of gutta-percha gum between, which are cemented together by steaming and pressing, using at all times a wet cloth under the hot iron. The brim is separate and distinct from the crown when made, and consists of a piece of thick wool padding, which is to be covered on both sides with the proper cloth, cemented together with the gum, first one side and then the other, after which the crown and brim are sewed together.

In all these cases, the gutta-percha gum acts not only as a cement but also a water-proofing to the hat.

Conclusion.

In this treatise upon the history of hats and hat-making, of furs, wools, &c., and the manufacture of felt, we are well aware of the impossibility of illustrating in full the hatting trade of America, as this country stands alone as compared with others, on account of the mixed population that is here collected. As we have representatives in this, as in every other line of business, from every civilized nation upon earth, with all their various methods of working in their own accustomed ways, the prejudices naturally engendered and entertained through habit being hard to combat, so that the judges of this work may be numerous and various, and no doubt profusely severe in some of their criticisms.

But there is going on a rapid amalgamation of all that is best in the trade of hatting, resulting from the continued flow of immigration, and heightened greatly by the wanderings of hatters generally, from shop to shop, and from town to town, that must ultimately bring together in this our beloved land, a perfection in the trade that cannot be attained by any other nation.

INDEX.

CATALOGUE

OF

PRACTICAL AND SCIENTIFIC BOOKS,

PUBLISHED BY

HENRY CAREY BAIRD,

INDUSTRIAL PUBLISHER,

No. 406 WALNUT STREET,

PHILADELPHIA.

☞ Any of the Books comprised in this Catalogue will be sent by mail, free of postage, at the publication price.

☞ This Catalogue will be sent, free of postage, to any one who will furnish the publisher with his address.

ARMENGAUD, AMOUROUX, AND JOHNSON.—THE PRACTICAL DRAUGHTSMAN'S BOOK OF INDUSTRIAL DESIGN, AND MACHINIST'S AND ENGINEER'S DRAWING COMPANION: Forming a complete course of Mechanical Engineering and Architectural Drawing. From the French of M. Armengaud the elder, Prof. of Design in the Conservatoire of Arts and Industry, Paris, and MM. Armengaud the younger and Amouroux, Civil Engineers. Rewritten and arranged, with additional matter and plates, selections from and examples of the most useful and generally employed mechanism of the day. By William Johnson, Assoc. Inst. C. E., Editor of "The Practical Mechanic's Journal." Illustrated by 50 folio steel plates and 50 wood-cuts. A new edition, 4to. . $10 00

ARROWSMITH.—PAPER-HANGER'S COMPANION: A Treatise in which the Practical Operations of the Trade are Systematically laid down: with Copious Directions Preparatory to Papering; Preventives against the Effect of Damp on Walls; the Various Cements and Pastes adapted to the Several Purposes of the Trade; Observations and Directions for the Panelling and Ornamenting of Rooms, &c. By James Arrowsmith, Author of "Analysis of Drapery," &c. 12mo, cloth $1 25

BAIRD.—THE AMERICAN COTTON SPINNER, AND MANAGER'S AND CARDER'S GUIDE:

A Practical Treatise on Cotton Spinning; giving the Dimensions and Speed of Machinery, Draught and Twist Calculations, etc.; with notices of recent Improvements: together with Rules and Examples for making changes in the sizes and numbers of Roving and Yarn. Compiled from the papers of the late ROBERT H. BAIRD. 12mo. $1 50

BAKER.—LONG-SPAN RAILWAY BRIDGES:

Comprising Investigations of the Comparative Theoretical and Practical Advantages of the various Adopted or Proposed Type Systems of Construction; with numerous Formulæ and Tables. By B. Baker. 12mo. $2 00

BAKEWELL.—A MANUAL OF ELECTRICITY—PRACTICAL AND THEORETICAL:

By F. C. BAKEWELL, Inventor of the Copying Telegraph. Second Edition. Revised and enlarged. Illustrated by numerous engravings. 12mo. Cloth $2 00

BEANS.—A TREATISE ON RAILROAD CURVES AND THE LOCATION OF RAILROADS:

By E. W. BEANS, C. E. 12mo. (In press.)

BLENKARN.—PRACTICAL SPECIFICATIONS OF WORKS EXECUTED IN ARCHITECTURE, CIVIL AND MECHANICAL ENGINEERING, AND IN ROAD MAKING AND SEWERING:

To which are added a series of practically useful Agreements and Reports. By JOHN BLENKARN. Illustrated by fifteen large folding plates. 8vo. $9 00

BLINN.—A PRACTICAL WORKSHOP COMPANION FOR TIN, SHEET-IRON, AND COPPER-PLATE WORKERS:

Containing Rules for Describing various kinds of Patterns used by Tin, Sheet-iron, and Copper-plate Workers; Practical Geometry; Mensuration of Surfaces and Solids; Tables of the Weight of Metals, Lead Pipe, etc.; Tables of Areas and Circumferences of Circles; Japans, Varnishes, Lackers, Cements, Compositions, etc. etc. By LEROY J. BLINN, Master Mechanic. With over One Hundred Illustrations. 12mo. $2 50

BOOTH.—MARBLE WORKER'S MANUAL:

Containing Practical Information respecting Marbles in general, their Cutting, Working, and Polishing; Veneering of Marble; Mosaics; Composition and Use of Artificial Marble, Stuccos, Cements, Receipts, Secrets, etc. etc. Translated from the French by M. L. BOOTH. With an Appendix concerning American Marbles. 12mo., cloth . . $1 50

BOOTH AND MORFIT.—THE ENCYCLOPEDIA OF CHEMISTRY, PRACTICAL AND THEORETICAL:

Embracing its application to the Arts, Metallurgy, Mineralogy, Geology, Medicine, and Pharmacy. By JAMES C. BOOTH, Melter and Refiner in the United States Mint, Professor of Applied Chemistry in the Franklin Institute, etc., assisted by CAMPBELL MORFIT, author of "Chemical Manipulations," etc. Seventh edition. Complete in one volume, royal 8vo., 978 pages, with numerous wood-cuts and other illustrations. $5 00

BOWDITCH.—ANALYSIS, TECHNICAL VALUATION, PURIFICATION, AND USE OF COAL GAS:

By Rev. W. R. BOWDITCH. Illustrated with wood engravings. 8vo. $6 50

BOX.—PRACTICAL HYDRAULICS:

A Series of Rules and Tables for the use of Engineers, etc. By THOMAS BOX. 12mo. $2 00

BUCKMASTER.—THE ELEMENTS OF MECHANICAL PHYSICS:

By J. C. BUCKMASTER, late Student in the Government School of Mines; Certified Teacher of Science by the Department of Science and Art; Examiner in Chemistry and Physics in the Royal College of Preceptors; and late Lecturer in Chemistry and Physics of the Royal Polytechnic Institute. Illustrated with numerous engravings. In one vol. 12mo. . $2 00

BULLOCK.—THE AMERICAN COTTAGE BUILDER:

A Series of Designs, Plans, and Specifications, from $200 to to $20,000 for Homes for the People; together with Warming, Ventilation, Drainage, Painting, and Landscape Gardening. By JOHN BULLOCK, Architect, Civil Engineer, Mechanician, and Editor of "The Rudiments of Architecture and Building," etc. Illustrated by 75 engravings. In one vol. 8vo. $3 50

BULLOCK.—THE RUDIMENTS OF ARCHITECTURE AND BUILDING:

For the use of Architects, Builders, Draughtsmen, Machinists, Engineers, and Mechanics. Edited by JOHN BULLOCK, author of "The American Cottage Builder." Illustrated by 250 engravings. In one volume 8vo. . . . $3 50

BURGH.—PRACTICAL ILLUSTRATIONS OF LAND AND MARINE ENGINES:

Showing in detail the Modern Improvements of High and Low Pressure, Surface Condensation, and Super-heating, together with Land and Marine Boilers. By N. P. BURGH, Engineer. Illustrated by twenty plates, double elephant folio, with text. $21 00

BURGH.—PRACTICAL RULES FOR THE PROPORTIONS OF MODERN ENGINES AND BOILERS FOR LAND AND MARINE PURPOSES.

By N. P. BURGH, Engineer. 12mo. . . . $2 00

BURGH.—THE SLIDE-VALVE PRACTICALLY CONSIDERED:

By N. P. BURGH, author of "A Treatise on Sugar Machinery," "Practical Illustrations of Land and Marine Engines," "A Pocket-Book of Practical Rules for Designing Land and Marine Engines, Boilers," etc. etc. etc. Completely illustrated. 12mo. $2 00

BYRN.—THE COMPLETE PRACTICAL BREWER:

Or, Plain, Accurate, and Thorough Instructions in the Art of Brewing Beer, Ale, Porter, including the Process of making Bavarian Beer, all the Small Beers, such as Root-beer, Ginger-pop, Sarsaparilla-beer, Mead, Spruce beer, etc. etc. Adapted to the use of Public Brewers and Private Families. By M. LA FAYETTE BYRN, M. D. With illustrations. 12mo. $1 25

BYRN.—THE COMPLETE PRACTICAL DISTILLER:

Comprising the most perfect and exact Theoretical and Practical Description of the Art of Distillation and Rectification; including all of the most recent improvements in distilling apparatus; instructions for preparing spirits from the numerous vegetables, fruits, etc.; directions for the distillation and preparation of all kinds of brandies and other spirits, spirituous and other compounds, etc. etc.; all of which is so simplified that it is adapted not only to the use of extensive distillers, but for every farmer, or others who may wish to engage in the art of distilling. By M. LA FAYETTE BYRN, M. D. With numerous engravings. In one volume, 12mo. $1 50

BYRNE.—POCKET BOOK FOR RAILROAD AND CIVIL ENGINEERS:

Containing New, Exact, and Concise Methods for Laying out Railroad Curves, Switches, Frog Angles and Crossings; the Staking out of work; Levelling; the Calculation of Cuttings; Embankments; Earth-work, etc. By OLIVER BYRNE. Illustrated, 18mo. $1 25

BYRNE.—THE HANDBOOK FOR THE ARTISAN, MECHANIC, AND ENGINEER:

By OLIVER BYRNE. Illustrated by 11 large plates and 185 Wood Engravings. 8vo. $5 00

BYRNE.—THE ESSENTIAL ELEMENTS OF PRACTICAL MECHANICS:

For Engineering Students, based on the Principle of Work. By OLIVER BYRNE. Illustrated by Numerous Wood Engravings, 12mo. $3 63

BYRNE.—THE PRACTICAL METAL-WORKER'S ASSISTANT:

Comprising Metallurgic Chemistry; the Arts of Working all Metals and Alloys; Forging of Iron and Steel; Hardening and Tempering; Melting and Mixing; Casting and Founding; Works in Sheet Metal; the Processes Dependent on the Ductility of the Metals; Soldering; and the most Improved Processes and Tools employed by Metal-Workers. With the Application of the Art of Electro-Metallurgy to Manufacturing Processes; collected from Original Sources, and from the Works of Holtzapffel, Bergeron, Leupold, Plumier, Napier, and others. By OLIVER BYRNE. A New, Revised, and improved Edition, with Additions by John Scoffern, M. B , William Clay, Wm. Fairbairn, F. R. S., and James Napier. With Five Hundred and Ninety-two Engravings; Illustrating every Branch of the Subject. In one volume, 8vo. 652 pages . $7 00

BYRNE.—THE PRACTICAL CALCULATOR:

For the Engineer, Mechanic, Manufacturer of Engine Work, Naval Architect, Miner, and Millwright. By OLIVER BYRNE. 1 volume, 8vo., nearly 600 pages $4 50

CABINET MAKER'S ALBUM OF FURNITURE:

Comprising a Collection of Designs for the Newest and Most Elegant Styles of Furniture. Illustrated by Forty eight Large and Beautifully Engraved Plates. In one volume, oblong $5 00

CALVERT.—LECTURES ON COAL-TAR COLORS, AND ON RECENT IMPROVEMENTS AND PROGRESS IN DYEING AND CALICO PRINTING:

Embodying Copious Notes taken at the last London International Exhibition, and *Illustrated with Numerous Patterns of Aniline and other Colors.* By F. GRACE CALVERT, F. R. S., F. C. S., Professor of Chemistry at the Royal Institution, Manchester, Corresponding Member of the Royal Academies of Turin and Rouen; of the Pharmaceutical Society of Paris; Sociêté Industrielle de Mulhouse, etc. In one volume, 8vo., cloth $1 50

CAMPIN.—A PRACTICAL TREATISE ON MECHANICAL ENGINEERING:

Comprising Metallurgy, Moulding, Casting, Forging, Tools, Workshop Machinery, Mechanical Manipulation, Manufacture of Steam-engines, etc. etc. With an Appendix on the Analysis of Iron and Iron Ores. By FRANCIS CAMPIN, C. E. To which are added, Observations on the Construction of Steam Boilers, and Remarks upon Furnaces used for Smoke Prevention; with a Chapter on Explosions. By R. Armstrong, C. E., and John Bourne. Rules for Calculating the Change Wheels for Screws on a Turning Lathe, and for a Wheel-cutting Machine. By J. LA NICCA. Management of Steel, including Forging, Hardening, Tempering, Annealing, Shrinking, and Expansion. And the Case-hardening of Iron. By G. EDE. 8vo. Illustrated with 29 plates and 100 wood engravings. $6 00

CAMPIN.—THE PRACTICE OF HAND-TURNING IN WOOD, IVORY, SHELL, ETC.:

With Instructions for Turning such works in Metal as may be required in the Practice of Turning Wood, Ivory, etc. Also, an Appendix on Ornamental Turning. By FRANCIS CAMPIN; with Numerous Illustrations, 12mo., cloth . . $3 00

CAPRON DE DOLE.—DUSSAUCE.—BLUES AND CARMINES OF INDIGO.

A Practical Treatise on the Fabrication of every Commercial Product derived from Indigo. By FELICIEN CAPRON DE DOLE. Translated, with important additions, by Professor H. DUSSAUCE. 12mo. $2 50

CAREY.—THE WORKS OF HENRY C. CAREY:

CONTRACTION OR EXPANSION? REPUDIATION OR RESUMPTION? Letters to Hon. Hugh McCulloch. 8vo. 38

FINANCIAL CRISES, their Causes and Effects. 8vo. paper 25

HARMONY OF INTERESTS; Agricultural, Manufacturing, and Commercial. 8vo., paper $1 00
Do. do. cloth $1 50

LETTERS TO THE PRESIDENT OF THE UNITED STATES. Paper 75

MANUAL OF SOCIAL SCIENCE. Condensed from Carey's "Principles of Social Science." By KATE McKEAN. 1 vol. 12mo. $2 25

MISCELLANEOUS WORKS: comprising "Harmony of Interests," "Money," "Letters to the President," "French and American Tariffs," "Financial Crises," "The Way to Outdo England without Fighting Her," "Resources of the Union," "The Public Debt," "Contraction or Expansion," "Review of the Decade 1857—'67," "Reconstruction," etc. etc. 1 vol. 8vo., cloth $4 50

MONEY: A LECTURE before the N. Y. Geographical and Statistical Society. 8vo., paper 25

PAST, PRESENT, AND FUTURE. 8vo. $2 50

PRINCIPLES OF SOCIAL SCIENCE. 3 volumes 8vo., cloth $10 00

REVIEW OF THE DECADE 1857—'67. 8vo., paper 38

RECONSTRUCTION: INDUSTRIAL, FINANCIAL, AND POLITICAL. Letters to the Hon. Henry Wilson, U. S. S. 8vo. paper 38

THE PUBLIC DEBT, LOCAL AND NATIONAL. How to provide for its discharge while lessening the burden of Taxation. Letter to David A. Wells, Esq., U. S. Revenue Commission. 8vo., paper 25

THE RESOURCES OF THE UNION. A Lecture read, Dec. 1865, before the American Geographical and Statistical Society, N. Y., and before the American Association for the Advancement of Social Science, Boston 25

THE SLAVE TRADE, DOMESTIC AND FOREIGN; Why it Exists, and How it may be Extinguished. 12mo., cloth $1 50

THE WAY TO OUTDO ENGLAND WITHOUT FIGHTING HER. Letters to the Hon. Schuyler Colfax, Speaker of the House of Representatives United States, on "The Paper Question," "The Farmer's Question," "The Iron Question," "The Railroad Question," and "The Currency Question." 8vo., paper 75

CHEVALIER.—THE PHOTOGRAPHIC STUDENT.

A Complete Treatise on the Theory and Practice of Photography. Translated from the French of A. CHEVALIER. Illustrated by numerous engravings. (In press.)

CLOUGH.—THE CONTRACTOR'S MANUAL AND BUILDER'S PRICE-BOOK:

Designed to elucidate the method of ascertaining, correctly, the value and Quantity of every description of Work and Materials used in the Art of Building, from their Prime Cost in any part of the United States, collected from extensive expericuce and observation in Building and Designing; to which are added a large variety of Tables, Memoranda, etc., indispensable to all engaged or concerned in erecting buildings of any kind. By A. B. CLOUGH, Architect, 24mo., cloth 75

COLBURN.—THE GAS-WORKS OF LONDON:

Comprising a sketch of the Gas-works of the city, Process of Manufacture, Quantity Produced, Cost, Profit, etc. By ZERAH COLBURN. 8vo., cloth 75

COLBURN.—THE LOCOMOTIVE ENGINE:

Including a Description of its Structure, Rules for Estimating its Capabilities, and Practical Observations on its Construction and Management. By ZERAH COLBURN. Illustrated. A new edition. 12mo. $1 25

COLBURN AND MAW.—THE WATER-WORKS OF LONDON:

Together with a Series of Articles on various other Waterworks. By ZERAH COLBURN and W. MAW. Reprinted from "Engineering." In one volume, 8vo. . . $1 00

DAGUERREOTYPIST AND PHOTOGRAPHER'S COMPANION:

12mo., cloth $1 25

DAVIS.—A TREATISE ON HARNESS, SADDLES, AND BRIDLES:

Their History and Manufacture from the Earliest Times down to the Present Period. By A. DAVIS, Practical Saddler and Harness Maker. (In press.)

DESSOYE.—STEEL, ITS MANUFACTURE, PROPERTIES, AND USE.

By J. B. J. Dessoye, Manufacturer of Steel; with an Introduction and Notes by Ed. Graten, Engineer of Mines. Translated from the French. In one volume, 12mo. (In press.)

DIRCKS.—PERPETUAL MOTION:

Or Search for Self-Motive Power during the 17th, 18th, and 19th centuries. Illustrated from various authentic sources in Papers, Essays, Letters, Paragraphs, and numerous Patent Specifications, with an Introductory Essay by Henry Dircks, C. E. Illustrated by numerous engravings of machines. 12mo., cloth $3 50

DIXON.—THE PRACTICAL MILLWRIGHT'S AND ENGINEER'S GUIDE:

Or Tables for Finding the Diameter and Power of Cogwheels; Diameter, Weight, and Power of Shafts; Diameter and Strength of Bolts, etc. etc. By Thomas Dixon. 12mo., cloth. $1 50

DUNCAN.—PRACTICAL SURVEYOR'S GUIDE:

Containing the necessary information to make any person, of common capacity, a finished land surveyor without the aid of a teacher. By Andrew Duncan. Illustrated. 12mo., cloth. $1 25

DUSSAUCE.—A NEW AND COMPLETE TREATISE ON THE ARTS OF TANNING, CURRYING, AND LEATHER DRESSING:

Comprising all the Discoveries and Improvements made in France, Great Britain, and the United States. Edited from Notes and Documents of Messrs. Sallerou, Grouvelle, Duval, Dessables, Labarraque, Payen, René, De Fontenelle, Malapeyre, etc. etc. By Prof. H. Dussauce, Chemist. Illustrated by 212 wood engravings. 8vo. $10 00

DUSSAUCE.—A GENERAL TREATISE ON THE MANUFACTURE OF EVERY DESCRIPTION OF SOAP:

Comprising the Chemistry of the Art, with Remarks on Alkalies, Saponifiable Fatty Bodies, the apparatus necessary in a Soap Factory, Practical Instructions on the manufacture of the various kinds of Soap, the assay of Soaps, etc. etc. Edited from notes of Larmé, Fontenelle, Malapeyre, Dufour, and others, with large and important additions by Professor H. Dussauce, Chemist. Illustrated. In one volume, 8vo. (In press.)

DUSSAUCE.—A PRACTICAL GUIDE FOR THE PERFUMER:

Being a New Treatise on Perfumery the most favorable to the Beauty without being injurious to the Health, comprising a Description of the substances used in Perfumery, the Formulæ of more than one thousand Preparations, such as Cosmetics, Perfumed Oils, Tooth Powders, Waters, Extracts, Tinctures, Infusions, Vinaigres, Essential Oils, Pastels, Creams, Soaps, and many new Hygienic Products not hitherto described. Edited from Notes and Documents of Messrs. Debay, Lunel, etc. With additions by Professor H. DUSSAUCE, Chemist. (In press, *shortly to be issued.*)

DUSSAUCE.—PRACTICAL TREATISE ON THE FABRICATION OF MATCHES, GUN COTTON, AND FULMINATING POWDERS.

By Professor H. DUSSAUCE. 12mo. $3 00

DUSSAUCE.—TREATISE ON THE COLORING MATTERS DERIVED FROM COAL TAR:

Their Practical Application in Dyeing Cotton, Wool, and Silk; the Principles of the Art of Dyeing and of the Distillation of Coal Tar, with a Description of the most Important New Dyes now in use. By Prof. H. DUSSAUCE. 12mo. . $3 00

DYER AND COLOR-MAKER'S COMPANION:

Containing upwards of two hundred Receipts for making Colors, on the most approved principles, for all the various styles and fabrics now in existence; with the Scouring Process, and plain Directions for Preparing, Washing-off, and Finishing the Goods. In one vol. 12mo. $1 25

EASTON.—A PRACTICAL TREATISE ON STREET OR HORSE-POWER RAILWAYS:

Their Location, Construction, and Management; with General Plans and Rules for their Organization and Operation; together with Examinations as to their Comparative Advantages over the Omnibus System, and Inquiries as to their Value for Investment; including Copies of Municipal Ordinances relating thereto. By ALEXANDER EASTON, C. E. Illustrated by 23 plates, 8vo., cloth $2 00

ERNI.—COAL OIL AND PETROLEUM:

Their Origin, History, Geology, and Chemistry; with a view of their importance in their bearing on National Industry. By Dr. HENRI ERNI, Chief Chemist, Department of Agriculture. 12mo. $2 50

ERNI.—THE THEORETICAL AND PRACTICAL CHEMISTRY OF FERMENTATION:

Comprising the Chemistry of Wine, Beer, Distilling of Liquors; with the Practical Methods of their Chemical Examination, Preservation, and Improvement—such as Gallizing of Wines. With an Appendix, containing well-tested Practical Rules and Receipts for the manufacture, etc., of all kinds of Alcoholic Liquors. By HENRY ERNI, Chief Chemist, Department of Agriculture. (In press.)

FAIRBAIRN.—THE PRINCIPLES OF MECHANISM AND MACHINERY OF TRANSMISSION:

Comprising the Principles of Mechanism, Wheels, and Pulleys, Strength and Proportions of Shafts, Couplings of Shafts, and Engaging and Disengaging Gear. By WILLIAM FAIRBAIRN, Esq., C. E., LL. D., F. R. S., F. G. S., Corresponding Member of the National Institute of France, and of the Royal Academy of Turin; Chevalier of the Legion of Honor, etc. etc. Beautifully illustrated by over 150 wood-cuts. In one volume 12mo. $2 50

FAIRBAIRN.—PRIME-MOVERS:

Comprising the Accumulation of Water-power; the Construction of Water-wheels and Turbines; the Properties of Steam; the Varieties of Steam-engines and Boilers and Wind-mills. By WILLIAM FAIRBAIRN, C. E., LL. D., F. R. S., F. G. S. Author of "Principles of Mechanism and the Machinery of Transmission." With Numerous Illustrations. In one volume. (In press.)

FLAMM.—A PRACTICAL GUIDE TO THE CONSTRUCTION OF ECONOMICAL HEATING APPLICATIONS FOR SOLID AND GASEOUS FUELS:

With the Application of Concentrated Heat, and on Waste Heat, for the Use of Engineers, Architects, Stove and Furnace Makers, Manufacturers of Fire Brick, Zinc, Porcelain, Glass, Earthenware, Steel, Chemical Products, Sugar Refiners, Metallurgists, and all others employing Heat. By M. PIERRE FLAMM, Manufacturer. Illustrated. Translated from the French. One volume, 12mo. (In press.)

GILBART.—A PRACTICAL TREATISE ON BANKING:

By JAMES WILLIAM GILBART. To which is added: THE NATIONAL BANK ACT AS NOW (1868) IN FORCE. 8vo. $4 50

GOTHIC ALBUM FOR CABINET MAKERS:
Comprising a Collection of Designs for Gothic Furniture. Illustrated by twenty-three large and beautifully engraved plates. Oblong $3 00

GRANT.—BEET-ROOT SUGAR AND CULTIVATION OF THE BEET:
By E. B. GRANT. 12mo. $1 25

GREGORY.—MATHEMATICS FOR PRACTICAL MEN:
Adapted to the Pursuits of Surveyors, Architects, Mechanics, and Civil Engineers. By OLINTHUS GREGORY. 8vo., plates, cloth $3 00

GRISWOLD.—RAILROAD ENGINEER'S POCKET COMPANION.
Comprising Rules for Calculating Deflection Distances and Angles, Tangential Distances and Angles, and all Necessary Tables for Engineers; also the art of Levelling from Preliminary Survey to the Construction of Railroads, intended Expressly for the Young Engineer, together with Numerous Valuable Rules and Examples. By W. GRISWOLD. 12mo., tucks. $1 25

GUETTIER.—METALLIC ALLOYS:
Being a Practical Guide to their Chemical and Physical Properties, their Preparation, Composition, and Uses. Translated from the French of A. GUETTIER, Engineer and Director of Founderies, author of "La Fouderie en France," etc. etc. By A. A. PESQUET, Chemist and Engineer. In one volume, 12mo. (In press, *shortly to be published.*)

HATS AND FELTING:
A Practical Treatise on their Manufacture. By a Practical Hatter. Illustrated by Drawings of Machinery, &c., 8vo.

HAY.—THE INTERIOR DECORATOR:
The Laws of Harmonious Coloring adapted to Interior Decorations: with a Practical Treatise on House-Painting. By D. R. HAY, House-Painter and Decorator. Illustrated by a Diagram of the Primary, Secondary, and Tertiary Colors. 12mo. $2 25

HUGHES.—AMERICAN MILLER AND MILLWRIGHT'S ASSISTANT:
By WM. CARTER HUGHES. A new edition. In one volume, 12mo. $1 50

HUNT.—THE PRACTICE OF PHOTOGRAPHY.

By Robert Hunt, Vice-President of the Photographic Society, London, with numerous illustrations. 12mo., cloth . 75

HURST.—A HAND-BOOK FOR ARCHITECTURAL SURVEYORS:

Comprising Formulæ useful in Designing Builder's work, Table of Weights, of the materials used in Building, Memoranda connected with Builders' work, Mensuration, the Practice of Builders' Measurement, Contracts of Labor, Valuation of Property, Summary of the Practice in Dilapidation, etc. etc. By J. F. Hurst, C. E. 2d edition, pocket-book form, full bound $2 50

JERVIS.—RAILWAY PROPERTY:

A Treatise on the Construction and Management of Railways; designed to afford useful knowledge, in the popular style, to the holders of this class of property; as well as Railway Managers, Officers, and Agents. By John B. Jervis, late Chief Engineer of the Hudson River Railroad, Croton Aqueduct, &c. One vol. 12mo., cloth $2 00

JOHNSON.—A REPORT TO THE NAVY DEPARTMENT OF THE UNITED STATES ON AMERICAN COALS:

Applicable to Steam Navigation and to other purposes. By Walter R. Johnson. With numerous illustrations. 607 pp. 8vo., half morocco $6 00

JOHNSON.—THE COAL TRADE OF BRITISH AMERICA:

With Researches on the Characters and Practical Values of American and Foreign Coals. By Walter R. Johnson, Civil and Mining Engineer and Chemist. 8vo. . . . $2 00

JOHNSTON.—INSTRUCTIONS FOR THE ANALYSIS OF SOILS, LIMESTONES, AND MANURES.

By J. W. F. Johnston. 12mo. 38

KEENE.—A HAND-BOOK OF PRACTICAL GAUGING,

For the Use of Beginners, to which is added A Chapter on Distillation, describing the process in operation at the Custom House for ascertaining the strength of wines. By James B. Keene, of H. M. Customs. 8vo. $1 25

KENTISH.—A TREATISE ON A BOX OF INSTRUMENTS,

And the Slide Rule; with the Theory of Trigonometry and Logarithms, including Practical Geometry, Surveying, Measuring of Timber, Cask and Malt Gauging, Heights, and Distances. By Thomas Kentish. In one volume. 12mo. . $1 25

KOBELL.—ERNI.—MINERALOGY SIMPLIFIED:

A short method of Determining and Classifying Minerals, by means of simple Chemical Experiments in the Wet Way. Translated from the last German Edition of F. VON KOBELL, with an Introduction to Blowpipe Analysis and other additions. By HENRI ERNI, M. D., Chief Chemist, Department of Agriculture, author of "Coal Oil and Petroleum." In one volume, 12mo. $2 50

LAFFINEUR.—A PRACTICAL GUIDE TO HYDRAULICS FOR TOWN AND COUNTRY;

Or a Complete Treatise on the Building of Conduits for Water for Cities, Towns, Farms, Country Residences, Workshops, etc. Comprising the means necessary for obtaining at all times abundant supplies of Drinkable Water. Translated from the French of M. JULES LAFFINEUR, C. E. Illustrated. (In press.)

LAFFINEUR.—A TREATISE ON THE CONSTRUCTION OF WATER-WHEELS:

Containing the various Systems in use with Practical Information on the Dimensions necessary for Shafts, Journals, Arms, etc., of Water-wheels, etc. etc. Translated from the French of M. JULES LAFFINEUR, C. E. Illustrated by numerous plates. (In press.)

LANDRIN.—A TREATISE ON STEEL:

Comprising the Theory, Metallurgy, Practical Working, Properties, and Use. Translated from the French of H. C. LANDRIN, Jr., C. E. By A. A. FESQUET, Chemist and Engineer. Illustrated. 12mo. (In press.)

LARKIN.—THE PRACTICAL BRASS AND IRON FOUNDER'S GUIDE:

A Concise Treatise on Brass Founding, Moulding, the Metals and their Alloys, etc.; to which are added Recent Improvements in the Manufacture of Iron, Steel by the Bessemer Process, etc. etc. By JAMES LARKIN, late Conductor of the Brass Foundry Department in Reany, Neafie & Co.'s Penn Works, Philadelphia. Fifth edition, revised, with Extensive additions. In one volume, 12mo. $2 25

LEAVITT.—FACTS ABOUT PEAT AS AN ARTICLE OF FUEL:
With Remarks upon its Origin and Composition, the Localities in which it is found, the Methods of Preparation and Manufacture, and the various Uses to which it is applicable; together with many other matters of Practical and Scientific Interest. To which is added a chapter on the Utilization of Coal Dust with Peat for the Production of an Excellent Fuel at Moderate Cost, especially adapted for Steam Service. By H. T. LEAVITT. Third edition. 12mo. $1 75

LEROUX.—A PRACTICAL TREATISE ON WOOLS AND WORSTEDS:
Translated from the French of CHARLES LEROUX, Mechanical Engineer, and Superintendent of a Spinning Mill. Illustrated by 12 large plates and 34 engravings. In one volume 8vo. (In press, *shortly to be published*)

LESLIE (MISS).—COMPLETE COOKERY:
Directions for Cookery in its Various Branches. By MISS LESLIE. 58th thousand. Thoroughly revised, with the addition of New Receipts. In 1 vol. 12mo., cloth . . $1 25

LESLIE (MISS). LADIES' HOUSE BOOK:
a Manual of Domestic Economy. 20th revised edition. 12mo., cloth $1 25

LESLIE (MISS).—TWO HUNDRED RECEIPTS IN FRENCH COOKERY.
12mo. 50

LIEBER.—ASSAYER'S GUIDE:
Or, Practical Directions to Assayers, Miners, and Smelters, for the Tests and Assays, by Heat and by Wet Processes, for the Ores of all the principal Metals, of Gold and Silver Coins and Alloys, and of Coal, etc. By OSCAR M. LIEBER. 12mo., cloth $1 25

LOVE.—THE ART OF DYEING, CLEANING, SCOURING, AND FINISHING:
On the most approved English and French methods; being Practical Instructions in Dyeing Silks, Woollens, and Cottons, Feathers, Chips, Straw, etc; Scouring and Cleaning Bed and Window Curtains, Carpets, Rugs, etc.; French and English Cleaning, any Color or Fabric of Silk, Satin, or Damask. By THOMAS LOVE, a Working Dyer and Scourer. In 1 vol. 12mo. $3 00

MAIN AND BROWN.—QUESTIONS ON SUBJECTS CONNECTED WITH THE MARINE STEAM-ENGINE:

And Examination Papers; with Hints for their Solution. By THOMAS J. MAIN, Professor of Mathematics, Royal Naval College, and THOMAS BROWN, Chief Engineer, R. N. 12mo., cloth $1 50

MAIN AND BROWN.—THE INDICATOR AND DYNAMOMETER:

With their Practical Applications to the Steam-Engine. By THOMAS J. MAIN, M. A. F. R., Ass't Prof. Royal Naval College, Portsmouth, and THOMAS BROWN, Assoc. Inst. C. E., Chief Engineer, R. N., attached to the R. N. College. Illustrated. From the Fourth London Edition. 8vo. . . . $1 50

MAIN AND BROWN.—THE MARINE STEAM-ENGINE.

By THOMAS J. MAIN, F. R. Ass't S. Mathematical Professor at Royal Naval College, and THOMAS BROWN, Assoc. Inst. C. E. Chief Engineer, R. N. Attached to the Royal Naval College. Authors of "Questions connected with the Marine Steam-Engine," and the "Indicator and Dynamometer." With numerous Illustrations. In one volume, 8vo. . . . $5 00

MAKINS.—A MANUAL OF METALLURGY:

More particularly of the Precious Metals: including the Methods of Assaying them. Illustrated by upwards of 50 Engravings. By GEORGE HOGARTH MAKINS, M. R. C. S., F. C. S., one of the Assayers to the Bank of England, Assayer to the Anglo-Mexican Mints, and Lecturer upon Metallurgy at the Dental Hospital, London. In one volume, 12mo. . . $3 50

MARTIN.—SCREW-CUTTING TABLES, FOR THE USE OF MECHANICAL ENGINEERS:

Showing the Proper Arrangement of Wheels for Cutting the Threads of Screws of any required Pitch; with a Table for Making the Universal Gas-Pipe Thread and Taps. By W. A. MARTIN, Engineer. 8vo. 50

MILES.—A PLAIN TREATISE ON HORSE-SHOEING.

With illustrations. By WILLIAM MILES, author of "The Horse's Foot," $1 00

MOLESWORTH. POCKET-BOOK OF USEFUL FORMULÆ AND MEMORANDA FOR CIVIL AND MECHANICAL ENGINEERS.

By GUILFORD L. MOLESWORTH, Member of the Institution of Civil Engineers, Chief Resident Engineer of the Ceylon Railway. Second American, from the Tenth London Edition. In one volume, full bound in pocket-book form . . $2 00

MOORE.—THE INVENTOR'S GUIDE:
Patent Office and Patent Laws; or, a Guide to Inventors, and a Book of Reference for Judges, Lawyers, Magistrates, and others. By J. G. MOORE. 12mo., cloth . . $1 25

MOREAU.—PRACTICAL GUIDE FOR THE JEWELLER,
In the Application of Harmony of Colors in the Arrangement of Precious Stones, Gold, etc., from the French of M. L. MOREAU, Jeweller and Designer. Illustrated. (In press.)

NAPIER.—CHEMISTRY APPLIED TO DYEING.
By JAMES NAPIER, F. C. S. A new and revised edition, brought down to the present condition of the Art. Illustrated. (In press.)

NAPIER.—A MANUAL OF DYEING RECEIPTS FOR GENERAL USE.
By JAMES NAPIER, F. C S. *With Numerous Patterns of Dyed Cloth and Silk.* Second edition, revised and enlarged. 12mo. $3 75

NAPIER.—MANUAL OF ELECTRO-METALLURGY:
Including the Application of the Art to Manufacturing Processes. By JAMES NAPIER. Fourth American, from the Fourth London edition, revised and enlarged. Illustrated by engravings. In one volume, 8vo. $2 00

NEWBERY.—GLEANINGS FROM ORNAMENTAL ART OF EVERY STYLE;
Drawn from Examples in the British, South Kensington, Indian, Crystal Palace, and other Museums, the Exhibitions of 1851 and 1862, and the best English and Foreign works. In a series of one hundred exquisitely drawn Plates, containing many hundred examples. By ROBERT NEWBERY. 4to. $15 00

NICHOLSON.—A MANUAL OF THE ART OF BOOK-BINDING:
Containing full instructions in the different Branches of Forwarding, Gilding, and Finishing. Also, the Art of Marbling Book-edges and Paper. By JAMES B. NICHOLSON. Illustrated. 12mo., cloth $2 25

NORRIS.—A HAND-BOOK FOR LOCOMOTIVE ENGINEERS AND MACHINISTS:
Comprising the Proportions and Calculations for Constructing Locomotives; Manner of Setting Valves; Tables of Squares, Cubes, Areas, etc. etc. By SEPTIMUS NORRIS, Civil and Mechanical Engineer. New edition. Illustrated, 12mo., cloth $2 00

NYSTROM.—ON TECHNOLOGICAL EDUCATION AND THE CONSTRUCTION OF SHIPS AND SCREW PROPELLERS:

For Naval and Marine Engineers. By JOHN W. NYSTROM, late Acting Chief Engineer U. S. N. Second edition, revised with additional matter. Illustrated by seven engravings. 12mo.

$2 50

O'NEILL.—CHEMISTRY OF CALICO PRINTING, DYEING, AND BLEACHING:

Including Silken, Woollen, and Mixed Goods; Practical and Theoretical. By CHARLES O'NEILL. (In press.)

O'NEILL.—A DICTIONARY OF CALICO PRINTING AND DYEING:

Containing a Brief Account of all the Substances and Processes in Use in the Arts of Printing and Dyeing Textile Fabrics; with Practical Receipts and Scientific Information. By CHARLES O'NEILL, Analytical Chemist, Fellow of the Chemical Society of London, etc. etc. Author of "Chemistry of Calico Printing and Dyeing." 8vo. (In press.)

OVERMAN—OSBORN.—THE MANUFACTURE OF IRON IN ALL ITS BRANCHES:

Including a Practical Description of the various Fuels and their Values, the Nature, Determination and Preparation of the Ore, the Erection and Management of Blast and other Furnaces, the characteristic results of Working by Charcoal, Coke, or Anthracite, the Conversion of the Crude into the various kinds of Wrought Iron, and the Methods adapted to this end. Also, a Description of Forge Hammers, Rolling Mills, Blast Engines, &c. &c. To which is added an Essay on the Manufacture of Steel. By FREDERICK OVERMAN, Mining Engineer. The whole thoroughly revised and enlarged, adapted to the latest Improvements and Discoveries, and the particular type of American Methods of Manufacture. With various new engravings illustrating the whole subject. By H. S. OSBORN, LL. D. Professor of Mining and Metallurgy in Lafayette College. In one volume, 8vo. (In press.) . $10 00

PAINTER, GILDER, AND VARNISHER'S COMPANION:

Containing Rules and Regulations in everything relating to the Arts of Painting, Gilding, Varnishing, and Glass Staining, with numerous useful and valuable Receipts; Tests for the Detection of Adulterations in Oils and Colors, and a statement of the Diseases and Accidents to which Painters, Gilders, and

Varnishers are particularly liable, with the simplest methods of Prevention and Remedy. With Directions for Graining. Marbling, Sign Writing, and Gilding on Glass. To which are added COMPLETE INSTRUCTIONS FOR COACH PAINTING AND VARNISHING. 12mo., cloth $1 50

PALLETT.—THE MILLER'S, MILLWRIGHT'S, AND ENGINEER'S GUIDE.

By HENRY PALLETT. Illustrated. In one vol. 12mo. $3 00

PERKINS.—GAS AND VENTILATION.

Practical Treatise on Gas and Ventilation. With Special Relation to Illuminating, Heating, and Cooking by Gas. Including Scientific Helps to Engineer-students and others. With illustrated Diagrams. By E. E. PERKINS. 12mo., cloth $1 25

PERKINS AND STOWE.—A NEW GUIDE TO THE SHEET-IRON AND BOILER PLATE ROLLER:

Containing a Series of Tables showing the Weight of Slabs and Piles to Produce Boiler Plates, and of the Weight of Piles and the Sizes of Bars to produce Sheet-iron; the Thickness of the Bar Gauge in Decimals; the Weight per foot, and the Thickness on the Bar or Wire Gauge of the fractional parts of an inch; the Weight per sheet, and the Thickness on the Wire Gauge of Sheet-iron of various dimensions to weigh 112 lbs. per bundle; and the conversion of Short Weight into Long Weight, and Long Weight into Short. Estimated and collected by G. H. PERKINS and J. G. STOWE $2 50

PHILLIPS AND DARLINGTON.—RECORDS OF MINING AND METALLURGY:

Or Facts and Memoranda for the use of the Mine Agent and Smelter. By J. ARTHUR PHILLIPS, Mining Engineer, Graduate of the Imperial School of Mines, France, etc., and JOHN DARLINGTON. Illustrated by numerous engravings. In one volume, 12mo. $2 00

PRADAL, MALEPEYRE, AND DUSSAUCE.—A COMPLETE TREATISE ON PERFUMERY:

Containing notices of the Raw Material used in the Art, and the Best Formulæ. According to the most approved Methods followed in France, England, and the United States. By M. P. PRADAL, Perfumer Chemist, and M. F. MALEPEYRE. Translated from the French, with extensive additions, by Professor H. DUSSAUCE. 8vo. $10 00

PROTEAUX.—PRACTICAL GUIDE FOR THE MANUFACTURE OF PAPER AND BOARDS.

By A. PROTEAUX, Civil Engineer, and Graduate of the School of Arts and Manufactures, Director of Thiers's Paper Mill, 'Puy-de-Dômé. With additions, by L. S. LE NORMAND. Translated from the French, with Notes, by HORATIO PAINE, A. B., M. D. To which is added a Chapter on the Manufacture of Paper from Wood in the United States, by HENRY T. BROWN, of the "American Artisan." Illustrated by six plates, containing Drawings of Raw Materials, Machinery, Plans of Paper-Mills, etc. etc. 8vo. $5 00

REGNAULT—ELEMENTS OF CHEMISTRY.

By M. V. REGNAULT. Translated from the French by T. FORREST BETTON, M. D., and edited, with notes, by JAMES C. BOOTH, Melter and Refiner U. S. Mint, and WM. L. FABER, Metallurgist and Mining Engineer. Illustrated by nearly 700 wood engravings. Comprising nearly 1500 pages. In two volumes, 8vo., cloth $10 00

SELLERS.—THE COLOR MIXER:

Containing nearly Four Hundred Receipts for Colors, Pastes, Acids, Pulps, Blue Vats, Liquors, etc. etc., for Cotton and Woollen Goods: including the celebrated Barrow Delaine Colors. By JOHN SELLERS, an experienced Practical Workman. In one volume, 12mo. $2 50

SHUNK—A PRACTICAL TREATISE ON RAILWAY CURVES AND LOCATION, FOR YOUNG ENGINEERS.

By WM. F. SHUNK, Civil Engineer. 12mo. . . . $1 50

SMEATON—BUILDER'S POCKET COMPANION:

Containing the Elements of Building, Surveying, and Architecture; with Practical Rules and Instructions connected with the subject. By A. C. SMEATON, Civil Engineer, etc. In one volume, 12mo. $1 25

SMITH.—THE DYER'S INSTRUCTOR:

Comprising Practical Instructions in the Art of Dyeing Silk, Cotton, Wool, and Worsted, and Woollen Goods: containing nearly 800 Receipts. To which is added a Treatise on the Art of Padding; and the Printing of Silk Warps, Skeins, and Handkerchiefs, and the various Mordants and Colors for the different styles of such work. By DAVID SMITH, Pattern Dyer. 12mo., cloth. $3 00

SMITH.—PARKS AND PLEASURE GROUNDS:

Or Practical Notes on Country Residences, Villas, Public Parks, and Gardens. By CHARLES H. J. SMITH, Landscape Gardener and Garden Architect, etc. etc. 12mo. . $2 25

STOKES.—CABINET-MAKER'S AND UPHOLSTERER'S COMPANION:

Comprising the Rudiments and Principles of Cabinet-making and Upholstery, with Familiar Instructions, Illustrated by Examples for attaining a Proficiency in the Art of Drawing, as applicable to Cabinet-work; The Processes of Veneering, Inlaying, and Buhl-work; the Art of Dyeing and Staining Wood, Bone, Tortoise Shell, etc. Directions for Lackering, Japanning, and Varnishing; to make French Polish; to prepare the Best Glues, Cements, and Compositions, and a number of Receipts particularly for workmen generally. By J. STOKES. In one vol. 12mo. With illustrations $1 25

STRENGTH AND OTHER PROPERTIES OF METALS.

Reports of Experiments on the Strength and other Properties of Metals for Cannon. With a Description of the Machines for Testing Metals, and of the Classification of Cannon in service. By Officers of the Ordnance Department U. S. Army. By authority of the Secretary of War. Illustrated by 25 large steel plates. In 1 vol. quarto $10 00

TABLES SHOWING THE WEIGHT OF ROUND, SQUARE, AND FLAT BAR IRON, STEEL, ETC.,

By Measurement. Cloth 63

TAYLOR.—STATISTICS OF COAL:

Including Mineral Bituminous Substances employed in Arts and Manufactures; with their Geographical, Geological, and Commercial Distribution and amount of Production and Consumption on the American Continent. With Incidental Statistics of the Iron Manufacture. By R. C. TAYLOR. Second edition, revised by S. S. HALDEMAN. Illustrated by five Maps and many wood engravings. 8vo., cloth $6 00

TEMPLETON.—THE PRACTICAL EXAMINATOR ON STEAM AND THE STEAM-ENGINE:

With Instructive References relative thereto, for the Use of Engineers, Students, and others. By WM. TEMPLETON, Engineer. 12mo. $1 25

THOMAS.—THE MODERN PRACTICE OF PHOTOGRAPHY.
By R. W. Thomas, F. C. S. 8vo., cloth . . . 75

THOMSON.—FREIGHT CHARGES CALCULATOR.
By Andrew Thomson, Freight Agent $1 25

TURNBULL.—THE ELECTRO-MAGNETIC TELEGRAPH:
With an Historical Account of its Rise, Progress, and Present Condition. Also, Practical Suggestions in regard to Insulation and Protection from the effects of Lightning. Together with an Appendix, containing several important Telegraphic Devices and Laws. By Lawrence Tnrnbull, M. D., Lecturer on Technical Chemistry at the Franklin Institute. Revised and improved. Illustrated. 8vo. $3 00

TURNER'S (THE) COMPANION:
Containing Instructions in Concentric, Elliptic, and Eccentric Turning; also various Plates of Chucks, Tools, and Instruments; and Directions for using the Eccentric Cutter, Drill, Vertical Cutter, and Circular Rest; with Patterns and Instruotions for working them. A new edition in one vol. 12mo. $1 50

ULRICH—DUSSAUCE.—A COMPLETE TREATISE ON THE ART OF DYEING COTTON AND WOOL:
As practised in Paris, Rouen, Mulhausen, and Germany. From the French of M. Louis Ulrich, a Practical Dyer in the principal Manufactories of Paris, Rouen, Mulhausen, etc. etc.; to which are added the most important Receipts for Dyeing Wool, as practised in the Manufacture Impériale des Gobelins, Paris. By Professor H. Dussauce. 12mo. $3 00

URBIN—BRULL.—A PRACTICAL GUIDE FOR PUDDLING IRON AND STEEL.
By Ed. Urbin, Engineer of Arts and Manufactures. A Prize Essay read before the Association of Engineers, Graduate of the School of Mines, of Liege, Belgium, at the Meeting of 1865—6. To which is added a Comparison of the Resisting Properties of Iron and Steel. By A. Brull. Translated from the French by A. A. Pesquet, Chemist and Engineer. In one volume, 8vo. $1 00

WATSON.—A MANUAL OF THE HAND-LATHE.
By Egbert P. Watson, Late of the "Scientific American," Author of "Modern Practice of American Machinists and Engineers." In one volume, 12mo. (In press.)

WATSON.—THE MODERN PRACTICE OF AMERICAN MACHINISTS AND ENGINEERS:

Including the Construction, Application, and Use of Drills, Lathe Tools, Cutters for Boring Cylinders, and Hollow Work Generally, with the most Economical Speed of the same, the Results verified by Actual Practice at the Lathe, the Vice, and on the Floor. Together with Workshop management, Economy of Manufacture, the Steam-Engine, Boilers, Gears, Belting, etc. etc. By EGBERT P. WATSON, late of the "Scientific American." Illustrated by eighty-six engravings. 12mo. . . $2 50

WATSON.—THE THEORY AND PRACTICE OF THE ART OF WEAVING BY HAND AND POWER:

With Calculations and Tables for the use of those connected with the Trade. By JOHN WATSON, Manufacturer and Practical Machine Maker. Illustrated by large drawings of the best Power-Looms. 8vo. $7 50

WEATHERLY.—TREATISE ON THE ART OF BOILING SUGAR, CRYSTALLIZING, LOZENGE-MAKING, COMFITS, GUM GOODS,

And other processes for Confectionery, &c. In which are explained, in an easy and familiar manner, the various Methods of Manufacturing every description of Raw and Refined sugar Goods, as sold by Confectioners and others . . $2 00

WILL.—TABLES FOR QUALITATIVE CHEMICAL ANALYSIS.

By Prof. HEINRICH WILL, of Giessen, Germany. Seventh edition. Translated by CHARLES F. HIMES, Ph. D., Professor of Natural Science, Dickinson College, Carlisle, Pa. . $1 25

WILLIAMS.—ON HEAT AND STEAM:

Embracing New Views of Vaporization, Condensation, and Expansion. By CHARLES WYE WILLIAMS, A. I. C. E. Illustrated. 8vo. $3 50

Spotted something that isn't quite right?

Let us know and we'll fix it for future reprints. And as a thank you for your diligence, we'll send you a free Sothis book of your choice.

publisher@sothis.press

This page intentionally left blank

Made in the USA
Monee, IL
08 November 2026

68106064R00059